STRESS CORROSION RESEARCH

NATO ADVANCED STUDY INSTITUTES SERIES

Proceedings of the Advanced Study Institute Programme, which aims at the dissemination of advanced knowlegde and the formation of contacts among scientists from different countries.

The series is published by an international board of publishers in conjunction with NATO Scientific Affairs Division

A	Life Sciences	Plenum Publishing Corporation
B	Physics	London and New York
C	Mathematical and Physical Sciences	D. Reidel Publishing Company Dordrecht and Boston
D	Behavioural and Social Sciences	Sijthoff & Noordhoff International Publishers B.V.
E	Applied Science	Alphen aan den Rijn, The Netherlands and Winchester, Mass., USA

Series E: Applied Science — No. 30

STRESS CORROSION RESEARCH

edited by

HANS ARUP
Korrosionscentralen ATV
Glostrup, Denmark

and

R. N. PARKINS
Department of Metallurgy and Engineering Materials
The University of Newcastle upon Tyne Newcastle upo Tyne, U.K.

SIJTHOFF & NOORDHOFF   1979
Alphen aan den Rijn — The Netherlands

Proceedings of the NATO Advanced Study Institute on Stress Corrosion
Research
Lyngby, Denmark
July 14-25, 1975

ISBN-13: 978-94-009-9573-4   e-ISBN-13: 978-94-009-9571-0
DOI: 10.1007/978-94-009-9571-0

TABLE OF CONTENTS

PREFACE

     Stress corrosion cracking, including hydrogen embrittlement,
has probably attracted more attention in the last decade than any
other single facet of the many that constitute the totality of
the environment sensitive behaviour of materials. To some this is
because the complex interactions between a number of parameters,
particularly metal composition and structure, electrochemistry
and the response of a metal to the application of stress, make
the subject of stress corrosion fascinating. To others it is because the subject has become increasingly important in practical
terms, as the problem of general corrosion has been controlled
and the borderline conditions between widespread attack and complete inactivity are more frequently encountered, and as materials
have become more efficiently used by operating at higher stress
levels than hitherto. Particularly in advanced engineering systems,
such as pressure vessels used in transportation or in the chemical
process industries or in the sophisticated equipment used in some
of the energy producing industries, the incidence of stress corrosion failure has increased alarmingly in recent times and with
consequences that are extremely costly if not worse.
     The reasons for holding a NATO Advanced Study Institute in
this field are therefore obvious, but why publish the proceedings
of the Institute in an age where there is, arguably, already a

superfluity of published material? Obe obvious answer is that the
papers presented constitute valuable reviews of detailed developments in recent times. Another is that major meetings of this type
frequently act as a watershed in relation to the establishment of
new concepts or to the abandonment of old ideas, and the meeting
in Copenhagen in 1975 may come to be regarded as assisting in this
crystallizing process.

Early work in the stress corrosion field perhaps overemphasized
the physico-metallurgical side of the problem at the expense of the
electro-chemical and stress aspects, but this imbalance has been
redressed in recent years. So far as electrochemistry is concerned,
although there is still much to be understood, we are now moving to
a situation in which failures in hitherto untried environments can
be anticipated by appropriate electrochemical measurement. The
application of the concepts of fracture mechanics has produced considerable benefits, especially with the higher strength or less
ductile materials, in relation to environment sensitive cracking,
but the Copenhagen ASI may well be remembered for the overall
emphasis that emerged upon the significance of crack tip strain rate.
The ramifications of the latter do not appear to be restricted to
the stress corrosion cracking of the more ductile materials, but
encompass hydrogen embrittlement and extend into the field of corrosion fatigue, the distinction of which from stress corrosion becomes
meaningless under some conditions. Some of these points were made in
the seminars that were held during the ASI, but others are reflected
in the published papers, and if subsequent research and technology
is thereby guided along more fruitful avenues the publication of the
present volume will be justified.

Finally this publication would not have been possible if
J.C. Scully, University of Leeds, Leeds, England, and H. Arup,
Corrosion Centre, Copenhagen, Denmark, had not spent so much of
their valuable time in organising the ASI, thereby creating the
conditions for the authors to prepare their papers.

R.N. Parkins

CONSTANT STRAIN RATE STRESS CORROSION TESTING

R.N. Parkins

INTRODUCTION

While there are a few laboratories in which strain rate stress
corrosion testing has been in use for some time, there remains a considerable amount of scepticism, and unfamiliarity with this method of
testing. A number of workers appear to have arrived at this method
of testing independently and experience in Newcastle in its use extends
over a period of some ten years, during which time some refinements to
the technique have been effected and there has evolved a better understanding of its implications. Like others who have come to use it,
we have found it increasingly useful, whilst recognising that there are
circumstances in which other stress corrosion test methods may be as
appropriate.

In essence the test involves the application of a relatively slow
strain rate ($\sim 10^{-6} \mathrm{sec}^{-1}$) to a specimen subjected to appropriate electrochemical conditions. Whilst the expression 'constant strain rate' has
become established by use, 'constant deflection rate' would be more
correct, since especially as cracks propagate the effective strain rate

will vary even for a constant deflection rate.   It should also be emphasized that the strain rates employed in these tests are very much slower
than those involved in straining electrode experiments of the type initially
performed by Hoar and West[1], and subsequently by various other workers,
and where the object of the experiments, to measure current transients,
is totally different.   In the measurements to which the present article
refers the object is to produce stress corrosion cracking that, metallographically, is indistinguishable from that produced in constant load or
constant strain experiments.   Nor is the test to be confused with that
adopted by Coleman et al[2] who used a constant rate of loading to determine the stress at which cracks were just detected when the specimen was
examined at X20.   Their specimens were loaded at a given rate (200 lb/
min on a 0.25" dia. specimen) to some arbitrary maximum stress and then
removed from the corrosion cell and examined, this cycle being repeated
until cracks were observed.   Our own use of constant strain rate testing
developed out of experiments designed to test some of the ideas put forward by Coleman et al in their paper, the disadvantages of constant rates
of loading soon being recognised and the constant strain rate type of
test was developed as a result[3].

In most laboratory corrosion experiments an attempt is made to
obtain data in a relatively short time, frequently by adopting some approach
that increases the severity of the test.   In stress corrosion testing this
usually takes the form of increasing the relative aggressiveness of the
environment by altering its composition, temperature, pressure, etc., by
increasing the relative susceptibility of the alloy through changes in
structure, by introducing a notch or pre-crack into the specimen or by
stimulating the corrosion reactions galvanostatically or potentiostatically.
The application of dynamic straining to a stress corrosion test specimen
comes into this category also, and like all the other accelerating approaches

to testing, its justification will vary according to the circumstances in
which it is used.   Stress corrosion crack velocities usually fall in the
range from 10^{-3} to 10^{-6} mm sec^{-1}, depending upon the system of metal and
environment involved, which imply that failure in laboratory test specimens
of usual dimensions should occur in not more than a few days.   This is
found to be so if the system is one in which stress corrosion cracks are
readily initiated, but it is common experience to find that many test pieces
do not fail in very extended periods of testing which then are usually
terminated at some arbitrarily selected time.   The consequences are that
considerable scatter may be associated with replicate tests and the arbitrary termination of the test leaves an element of doubt concerning the
outcome if it had been allowed to continue to a longer time.   Just as
the use of pre-cracked specimens assists in stress corrosion crack initiation,
so apparently does the application of slow dynamic strain, which has the
further advantage that the test is not terminated after some arbitrary time,
since the conclusion is always achieved by the specimen fracturing and the
criterion of cracking susceptibility is then related to the mode of fracture.
Thus, in the form in which it is normally employed the slow strain rate
method will usually result in failure in not more than about two days,
either by ductile fracture or by stress corrosion cracking, according to
the susceptibility towards the latter, and metallographic or other parameters
may then be assigned in assessing the cracking response.   The fact that the
test concludes in this positive manner in a relatively short period of time
constitutes one of its main attractions.

Early use of the test was in providing data whereby the effects
of such variables as alloy composition and structure or inhibitive additions
to cracking environments could be compared, and also for promoting stress
corrosion cracking in combinations of alloy and environment that could not
be caused to fail in the laboratory under conditions of constant load or

constant strain. Thus, they constitute a relatively severe type of test in the sense that they frequently promote stress corrosion failure in the laboratory where other modes of stressing plain specimens do not promote cracking, and in this sense they are in a similar category to tests on pre-cracked specimens.

In recent years understanding of the implications of dynamic strain testing has developed and it now appears that this type of test may have much more relevance and significance than just that of an effective, rapid, sorting test. It may, at first sight, be argued that laboratory tests involving the pulling of specimens to failure at a slow strain rate shows little relation to the reality of service failures. In point of fact in constant strain and constant load tests crack propagation also occurs under conditions of slow dynamic strain to a greater or lesser extent depending upon the initial value of stress, the point in time during the test at which a stress corrosion crack is initiated and various metallurgical parameters that govern creep in the specimen. Moreover, there is an increasing amount of evidence for some systems which suggests that the function of stress in stress corrosion cracking is to promote a strain rate which, rather than stress per se, is the parameter that really governs crack initiation or propagation. In these cases the minimum creep rate for cracking is as much an engineering design parameter as is the threshold stress or stress intensity obtained from constant load tests on plain or pre-cracked specimens.

THE TEST METHOD

The form of the apparatus that we use most frequently is shown in Figure 1 and consists of a moderately stiff (hard) straining frame and a drive mechanism that produces six crosshead speeds in the range from about 10^{-4} to 10^{-7} in sec^{-1}. Whilst a variable speed motor could be employed to give an appropriate range of strain rates, we have found that a constant speed motor driving through reduction boxes, gear wheels and chain and sprocket drives gives reliable positive drive that is readily varied by gear changes.

Figure 2 shows the straining frame in some detail. The outer frame, A, incorporating a 5000 lb. load cell, B, has a sub-frame assembly, C, consisting of two moving crossheads and tie-bolts, that moves vertically over the outer frame on accurately aligned bushes, D. The test specimen, surrounded by the corrosion cell, is contained by grips within the moving sub-frame, being attached to the bottom crosshead of the latter through the linkage, E, and through the linkage, F, and the load cell to the top, fixed, crosshead of the outer frame. Hemispherical seatings at the points of fixing, G and H, and in the grips and various pinned joints ensure axial alignment of the specimen, with adequate clearance in a hole through the top crosshead of the sub-frame to allow the load to be transmitted from specimen to load cell without interference upon the link rod (F). The moving sub-frame is counterbalanced by helical springs, L, and is caused to move by the rotation of the lead screw, M, having 20 threads to the inch, that has its mating thread in N, the whole of this rotating part being supported by appropriate bearings housed in O. The rotation of the lead screw is effected by a worm, mounted on the shaft P, and wheel, Q. The worm and wheel may be disengaged to allow rapid adjustment of the position of the sub-frame at the beginning of a test.

The drive mechanism is shown schematically in Figure 3 and consists of a $\frac{1}{4}$ h.p. motor, R, driving through two 60:1 reduction boxes, S, through flexible couplings. From the second reduction box, a chain drive, T, joining a 17 tooth to a 19 tooth sprocket, rotates a bearing mounted shaft, U, upon which there are three spur gears having 75, 45 and 15 teeth respectively, any one of which may be mated with its appropriate gear on a bearing mounted second shaft, V, the mating gears having 15, 47 and 75 teeth respectively.   From this variable gear box, a further chain drive, W, joins a 17 tooth sprocket to either a 19 or 76 tooth sprocket mounted on the same shaft, P, as the worm on the straining frame.

Although a variety of specimen shapes and sizes, including pre-cracked specimens, have been used with these machines, the specimen most frequently employed is a plain cylindrical test piece some 6-7" long and 3/16" dia., threaded at each end for insertion into the grips.   A gauge length, incorporating a parallel portion 0.5" to 1" long, and in which the diameter is reduced to 0.1", is standard.   The corrosion cell may be as simple as a glass tube, closed at each end by rubber stoppers through which the specimen passes, as in Figure 1, or, if the environment requires it, a more sophisticated arrangement made from Teflon and incor-porating a heating element and items that permit testing with electrochemical control.   Deegan and Wilde[4] have described a test chamber for use with liquid ammonia that permits dynamic straining of the specimen whilst the pressure is maintained at 125 psig.

ASSESSMENT OF RESULTS FROM DYNAMIC STRAIN TESTS

The usual means of indicating the severity of stress corrosion cracking in a static test on a plain specimen is the time to failure at a given stress or the threshold stress if tests are conducted over a range of initial stresses.   At first sight neither of these parameters would appear appropriate for assessing cracking severity in constant strain rate tests.   Metallographic examination can confirm whether or not stress corrosion cracking has occurred during a dynamic strain test, although this does not usually lead to a quantifiable result for purposes of comparison.   (We have employed maximum crack depth determined on speci-mens tested for a given time or to an arbitrary strain, insufficient to achieve the U.T.S., in a dynamic test, but a considerable amount of tedious metallography may be involved in such measurements.)   An examination of the test soon indicates that a number of readily measurable and quantifiable parameters are available for assessing cracking severity in constant strain rate tests, arising out of the fact that stress corrosion failures are usually associated with very little plastic deformation during crack propagation.   Thus, measures of ductility, such as percent reduction in area or elongation, may be used producing results of the form shown in Figure 4 for the cracking of a mild steel in a boiling NaOH solution at various controlled potentials[5].   Metallographic examination of specimens from such tests would reveal intergranular stress corrosion cracks associated with the low ductilities and transgranular failures, with dimpled fracture surfaces, associated with the high ductility values.   The effect of stress corrosion cracking upon the observed ductility is, of course, reflected in the stress-strain curves that may be obtained by continuous recording of the response from the load cell, Figure 5 showing the forms of curves obtained with and without attendant stress corrosion.   It will be apparent from this Figure that not only is the strain to fracture dependent upon the presence or otherwise of stress corrosion cracks, but so also is the maximum load achieved.   The latter may be used with convenience for expressing cracking susceptibility in some systems, the variation of this with strain rate in some tests on a mild steel in a boiling nitrate solution[3], as shown in Figure 6, indicating the justification for such an approach.   However, the variations in maximum load achieved in constant strain rate tests in circumstances of varying cracking severity are not always large enough for significant distinctions to be made and even measurements of ductility, such as reduction in area, are not invariably easily made, if only because the final fracture of the specimen does not always follow a simple path and the fitting of the two broken

pieces together is not easy.   In such cases a combination of load and
ductility may provide a useful basis of comparison, since the area under
the stress-strain curve, as is apparent from Figure 5, can be used for
assessing cracking susceptibility[6].   However, probably the easiest
quantity to measure with reasonable accuracy in a constant strain rate
test is the time to failure, since this will correspond to the load
returning to zero when the specimen finally fractures, a feature easily
determined from a continuous recording of the response from the load cell.
That the time to failure may be used in comparing cracking tendencies is
apparent from Figure 7, which shows that the time to failure and the % R.A.
are simply related to one another, a result that is hardly surprising when
it is remembered that the lesser the intensity of stress corrosion cracking
the greater will be the ductility to fracture and, therefore, the greater
the time to fracture for a constant strain rate.   In our work we now most
frequently use the time to failure as a measure of cracking susceptibility,
the results usually being normalized by dividing by the time to failure
at the same strain rate in a test in an inert environment at the same
temperature as that employed in the stress corrosion test, so that
increasing susceptibility is marked by increasing departure of this
ratio from unity.

CHOICE OF STRAIN RATE

It is most important to appreciate that the same strain rate does
not produce the same response in all systems.   Clearly if the strain rate
is too high ductile fracture by void coalescence will occur before the
necessary corrosion reactions can take place, as is apparent from Figures
4 and 6, and consequently relatively slow strain rates, usually in the
region of 10^{-5} to $10^{-6} sec^{-1}$, depending upon the system, need to be employed
to produce stress corrosion.   However, it is possible in some systems for

the strain rate to be too low for stress corrosion cracking to be produced. Thus, it is conceivable that if the environment is one that produces filming of the metal surface, and probably almost every stress corrosion environment does, then at very slow strain rates film repair may be fast enough to keep pace with the rate at which bare metal is formed, so that the cracking reaction is not sustained. It may be expected therefore, that if the strain rate is varied over a wide enough range the cracking susceptibility will be found to pass through a maximum with reduced susceptibility at both higher and lower strain rates. The results shown in Figure 8 indicate such an effect for a Mg-Al alloy immersed in a chloride-chromate solution[7], the results also indicating that the range of strain rates within which stress corrosion cracking occurs is dependent upon the composition of the environment, the range being extended as the environment is made more aggressive, i.e. has a lesser filming tendency, by an increase in the chloride/chromate ratio, as would be expected. The critical strain rate range is also dependent upon the polarity and magnitude of applied currents for constant solution composition[8], as is apparent from Figure 9. Scully and Powell[9] have published a curve suggesting a similar effect for a Ti-Al-Sn alloy immersed in 3% NaCl solution.

There are various indications suggesting that the strain rate effects mentioned above can be operative in constant strain or constant load stress corrosion tests. In such tests the metal creeps at a diminishing rate due to creep exhaustion after the stress is applied, unless crack propagation results in an effective increase in stress that accelerates creep. It may be expected therefore that the stress corrosion cracking response will depend upon the delay between the application of stress and the bringing of the environment into contact with the test specimen. Figure 10 shows the results[7] obtained on a Mg-Al alloy tested in a chloride-chromate solution in which in one series the specimens

were stressed after being immersed in the environment and in the other
series were allowed to undergo creep relaxation for 3 hr in the test
cell after loading before the solution was introduced into the cell.
Clearly prior creep raises the threshold stress and the times to failure
above the latter vary in a manner that may be expected in the light of the
results given earlier.

Results from a totally different system, of a C-Mn steel immersed
in a CO_3-HCO_3 solution, support those given above. In these tests fatigue
pre-cracked specimens were used and were loaded as cantilevers in the equipment
fatigue pre-crack, despite the fact that the stress at the crack tip was
essentially the same as that which had produced cracking in the earlier
test (Curve A). One obvious explanation of this result was that by the
time the potential was changed to one that should promote cracking the
creep rate had diminished to a value below that which sustains cracking.
A further experiment was therefore conducted in which the potential was
changed from -950 to -650 mV before the creep rate had fallen to such a
low value as that which obtained in Curve C and, as Curve D indicates,
this resulted in a stress corrosion crack being propagated, as was
confirmed by metallographic examination. This experiment suggested
that if, instead of constant load tests, the specimens were subjected to
various beam deflection rates it should be possible to define a minimum
deflection rate below which cracking is not observed. A series of tests
was conducted in the equipment shown in Figure 11 in which each specimen
was subjected to a different beam deflection for a period of five days,
all of the specimens being initially deflected to produce effectively the
same preload. (Strain gauges attached to the beam and previously cali-
brated allowed the equivalent load upon the specimen to be determined at
any time; the load changes in the various beam deflection rate tests
were not more than a few per cent.) Metallographic examination of the

specimens after testing allowed the length of the stress corrosion crack
to be determined and, using the time of the test, a crack velocity could
be determined for each experiment.   The results of a typical series of
tests are shown in Figure 13 and clearly indicate a lower limiting strain
rate below which crack propagation is not observed and above which the
crack velocity remains essentially constant irrespective of strain rate
shown in Figure 11, the pre-cracked specimens being employed as a convenient
way of locating the region in which the intergranular stress corrosion cracks
occur.   Cracking in this system only occurs within a relatively narrow range
of potentials and if the load is applied to the specimen after the potential
has been established in the cracking range the deflection of the beam,
measured by a displacement transducer, reflects occurrences in the region
of the crack tip in the manner shown in Curve A in Figure 12.   The interpretation of the shape of this curve is that, initially the beam deflection
rate diminishes at a relatively rapid rate as the creep in the plastic zone
at the crack tip becomes exhausted, but after a stress corrosion crack has
been initiated and propagated to some extent the constant load condition
leads to an increase in stress and additional creep, so that the beam
deflection rate begins to accelerate under the joint actions of crack
propagation and creep in the metal beyond the crack tip.   Now if the
potential is not held in the stress corrosion cracking range, the beam
deflection curve simply reflects the creep behaviour of the specimen after
loading and is shown as Curve B in Figure 12, so that stress corrosion
crack propagation is readily detected in the differences between Curves
A and B.   If an experiment is now conducted in which the potential is
initially held outside the cracking range at -950 mV for about 1 day after
the load is applied and then the potential is moved to -650 mV, i.e. inside
the cracking range, the beam deflection curve (C in Figure 12) suggests
that stress corrosion cracking has not occurred, since it is essentially

the same as that obtained when the potential is maintained at -950 mV throughout. Metallographic examination confirmed that no intergranular stress corrosion crack had propagated from the tip of the transgranular over the range studied; in fact at even higher beam deflection rates than those indicated the crack velocity may increase beyond the plateau region, possibly with a change in cracking mode from intergranular to transgranular. However, more importantly from the present viewpoint, the minimum strain rate for cracking shown in Figure 13 is found to be dependent upon such parameters as potential and temperature and is further evidence of the necessity for choosing an appropriate strain rate for the particular system being studied by dynamic strain testing.

This implication of achieving a critical balance between the film growth rate and the rate at which bare metal is created by straining if cracking is to ensue means, since the filming rates observed on different metals in different environments are variable, that the appropriate strain rate for a particular system should be determined for each case. This implies that tests should be conducted at different strain rates and our experiences with a wide variety of systems of metal and environment are that cracking has always been produced by strain rates somewhere in the range from 10^{-4} to 10^{-6} mm sec^{-1}; systems that undergo rapid film growth, such as those involving alloys based upon Ti or Mg, usually require the higher strain rates, whilst alloys based upon Cu or Fe usually respond more appropriately at the slower end of this range.

SOME RESULTS FROM DYNAMIC STRAIN TESTS

Notwithstanding the possible mechanistic implications of strain rate testing to stress corrosion, it is likely that the method has considerable potential for routine testing where a rapid assessment of alloy or environmental effects is required. Below examples are given of the uses that we have found for the test in our own work.

Clearly for the test to have credence it should give results that are comparable with those obtained by other methods. To illustrate the extent to which it achieves this end, Figure 14 shows some results for tests upon some low alloy steels in boiling $4N$ NH_4NO_3. Various alloying elements were added to these the effect of which was to produce a range of cracking susceptibilities as measured by the threshold stress determined from a series of constant strain tests made with different initial stresses. To allow for the influence of the alloying upon the mechanical properties of the steels the stress corrosion test results have been normalized by dividing the threshold stress for each steel by its yield strength. The same steels were also tested in dynamic strain tests in $4N$ NH_4NO_3 and in silicone oil at the same temperature, so that the cracking susceptibility may be expressed as a time to failure ratio. Whilst the results shown in Figure 14 indicate some scatter, the general trend is clear in that the results from the two types of test show reasonable agreement in placing the steels in essentially the same order of merit. This is particularly so when the differences are relatively large, whilst the scatter or discrepancies at similar susceptibilities is probably related to the fact that in the constant strain tests steels with identical threshold stresses may show different stress-time to failure curves at higher stresses. In dynamic strain tests it appears likely that stress corrosion crack initiation will occur when the stress achieves some value in the region of the threshold stress, but thereafter the stress will continue to increase in a manner related to the crack propagation rate and is likely to achieve values significantly above the threshold stress observed in a constant strain test. If the times to failure in the latter at stresses above the threshold are prolonged because of a slow crack propagation rate, then in strain rate tests a longer lifetime is to be expected and this is probably the reason why the three steels shown in Figure 14 as having values of σ_{th}/σ_{YS} very close to unity exhibit somewhat different time to failure ratios.

The reflection of crack velocities in strain rate test results is apparent in data from such tests upon a carbon steel in three very different environments, all of which promote intergranular failure. Figure 15 shows the time to failure ratios from strain rate tests upon a 0.07% C steel in a nitrate and a hydroxide solution, both at their boiling points, and in a carbonate-bicarbonate solution at 90°C, at various controlled potentials. Not only do these solutions promote stress corrosion cracking in different ranges of potential, but they also promote very different crack velocities. Thus, at the potentials that are associated with maximum cracking suscepti-bility the crack velocities in the three environments are : NO_3 - 2.75 x 10^{-4} mm sec^{-1}; OH - 1.9 x 10^{-5} mm sec^{-1}; CO_3-HCO_3 - 2.5 x 10^{-6} mm sec^{-1}. Figure 15 indicates that the time to failure ratios at the potentials which produce maximum cracking susceptibility in the three different environments are inversely dependent upon the measured crack velocities.

An example[10] of the use of strain rate testing in comparing the effects of structural variations in a C-Mn steel upon stress corrosion cracking propensities provides additional evidence of the correspondence between results from this method of testing and that involving constant strain. Figure 16 shows the applied stress-time to failure curves for a 0.23 C steel after water quenching and tempering for 1 hr at various temperatures, together with the curve for the steel in the annealed condition, the stress corrosion test environment being a boiling nitrate solution. Clearly quenching and tempering at temperatures of 400°C and below increase susceptibility to cracking as compared with the response in the annealed condition, whilst tempering at 600°C improves cracking resistance, as measured in terms of threshold stresses or the times to failure at a stress such as 450 N/mm^2.   Figure 17 shows the time to

failure ratios from constant strain rate tests for an almost identical

steel tested in the same solution and this demonstrates the same trends

with respect to the effects of tempering temperature as were shown by

constant strain tests in that low temperatures increase susceptibility

and high temperatures decrease cracking response as compared with that

observed with the annealed structure.   It is also worthy of mention

that, whereas the data shown in Figure 16 took about six weeks to compile,

the results shown in Figure 17 were obtained in less than one week, using

a single test facility in each case.

    The speed with which results can be obtained by strain rate

testing in contrast to the situation in more conventional tests is brought

out in dramatic fashion in the work of Deegan and Wilde[4], who used dynamic

strain testing, and the earlier comparable work on the stress corrosion crack-

ing of a low alloy steel in liquid ammonia by Loginow and Phelps[11].   In

the latter work, in which constant strain tuning fork and tensile specimens

were used, test times were of the order of 1-3 years with multiple tests

necessary because the frequency of failure was in the region of 20%.   On

the other hand, Deegan and Wilde were able to confirm the conclusions of

Loginow and Phelps, tests being conducted "in duplicate or as many times

as was necessary to achieve a repeatability of $\pm$ 10%", and failures

occurring in not more than 2 days.

    This facility of strain rate testing for comparing environmental

effects in relation to stress corrosion cracking, and in producing results

in a relatively short time, constitutes one of its major attractions.   The

demonstration[12] of carbonate-bicarbonate solutions as stress corrosion

cracking media for C steels was, in part, achieved only because of the

use of strain rate testing.   One final example of the use of the method

in this particular area will suffice to underline its attractions, namely in the assessment of additions to caustic solutions to prevent the stress corrosion of mild steels[5]. Figure 18 shows the effects of various additions to boiling 35% NaOH upon the reduction in area of a mild steel tested at various controlled potentials in a constant strain rate test and indicates that PbO, Pb_3O_4 and ZnO are without effect, that Na_2SiO_3 has a partially beneficial effect, whilst the tannins, quebracho and valonea, and a phosphate are good inhibitors of caustic cracking. Clearly the same conclusion could have been obtained by testing in each solution at the single potential -950 mV. However, the dangers of performing tests at a single potential will be apparent from Figure 19, which shows that none of the additions $KMnO_4$, $NaNO_3$ and Na_2SO_4 has any significant inhibitive effect, although they do cause the potential range for cracking to be moved. The effect of the latter is such that if tests had been conducted at the single potential -950 mV, the conclusion may well have been that $KMnO_4$ additions, at least, have a beneficial effect. This result may be interpreted as supporting the use of potentiostatic control in stress corrosion tests, which it does, rather than showing any advantage of constant strain rate tests. However, in constant load or constant strain tests on the same steel in the same solutions no failures whatsoever could be obtained, even with potential control, except in constant load tests when the initial applied stress was in the region of the U.T.S. of the steel and which therefore virtually amounted to dynamic strain tests in view of the instability associated with stressing to the U.T.S.

REFERENCES

1. T.P. Hoar and J.M. West, Proc. Roy. Soc., A268, 304, 1962.

2. E.G. Coleman, D. Weinstein and W. Rostoker, Acta. Met., 9, 491, 1961.

3. M. Henthorne, Ph.D. Thesis, University of Newcastle upon Tyne, England, 1965.

4. D.C. Deegan and B.E. Wilde, International Conference on Stress Corrosion Cracking and Hydrogen Embrittlement of Iron Base Alloys, Unieux-Firminy, France, June 1973, to be published by NACE.

5. M.J. Humphries and R.N. Parkins, Corr. Sci., 7, 747, 1967.

6. J. Flis and J.C. Scully, Corrosion 24, 326, 1968.

7. W.R. Wearmouth, G.P. Dean and R.N. Parkins, Corrosion, 29, 251, 1973.

8. G.P. Dean, Ph.D. Thesis, University of Newcastle upon Tyne, England, 1971

9. J.C. Scully and D.T. Powell, Corr. Sci., 10, 719, 1970.

10. R.N. Parkins, P.W. Slattery, W.R. Middleton and M.J. Humphries, Brit. Corr. Jnl., 8, 117, 1973.

11. A.W. Loginow and E.N. Phelps, Corrosion, 18, 299, 1962.

12. J.M. Sutcliffe, R.R. Fessler, W.K. Boyd and R.N. Parkins, Corrosion, 28, 313, 1972.

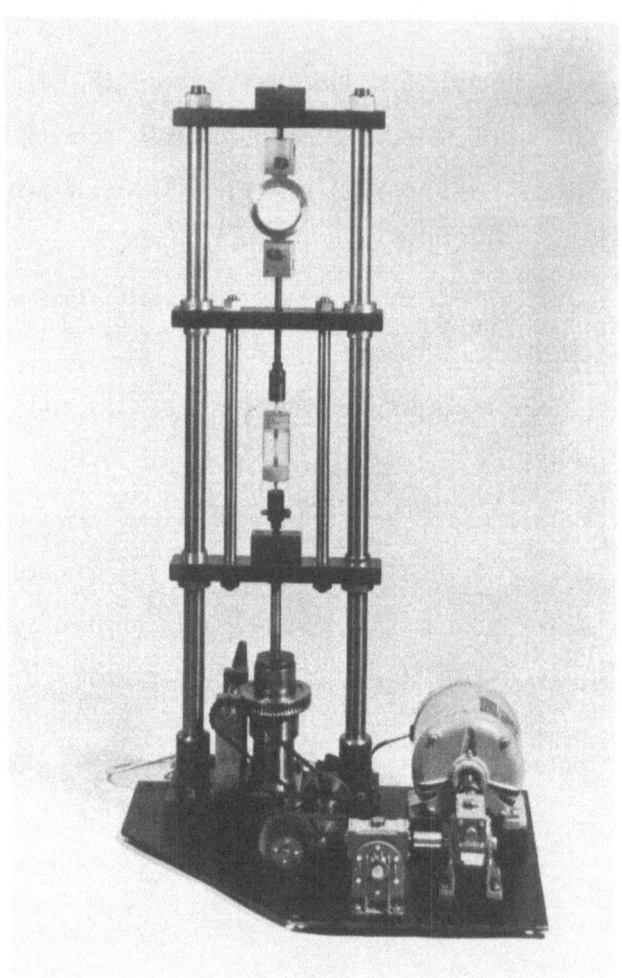

FIGURE 1.   Testing machine for constant
deflection rate tensile tests.

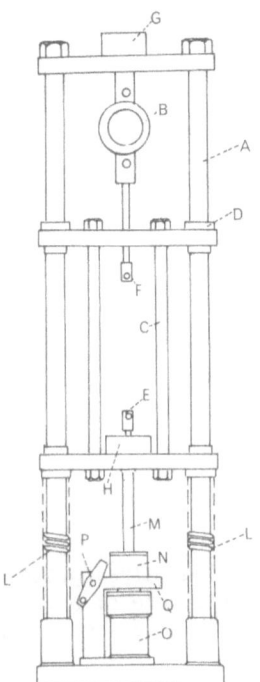

FIGURE 2.   Components of frame of constant
deflection rate testing machine.

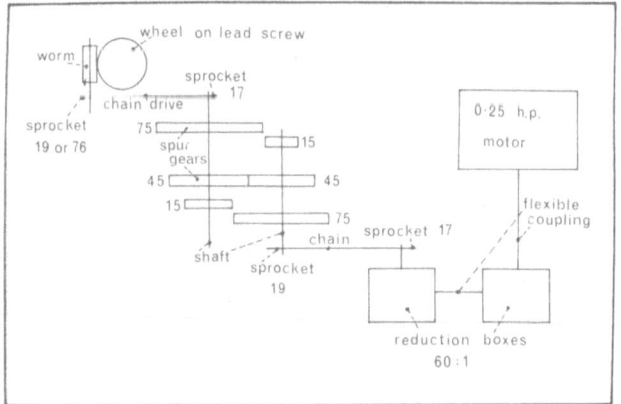

FIGURE 3.   Schematic of drive unit for constant
deflection rate testing machine.

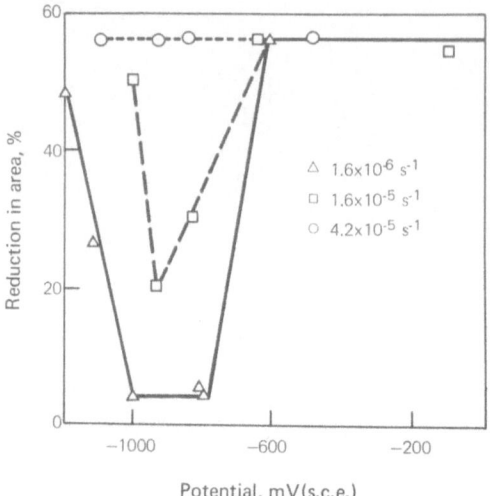

FIGURE 4.   Effects of various strain rates upon the
            ductility to fracture of a C steel in
            boiling 35% NaOH at different controlled
            potentials.

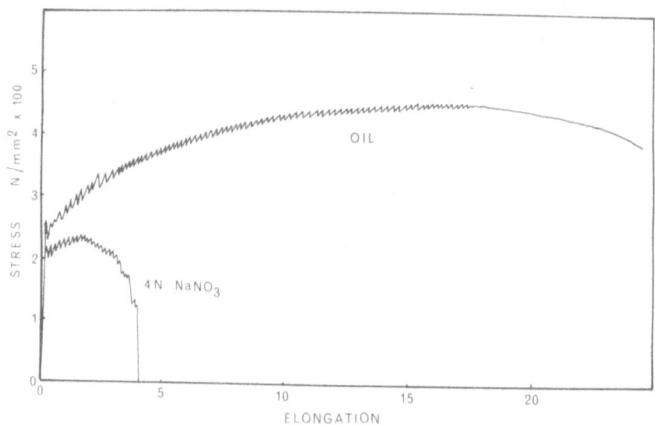

FIGURE 5.   Stress-elongation curves for C steel in
            constant deflection rate test in boiling
            4N NaNO$_3$ and in oil at the same temperature.

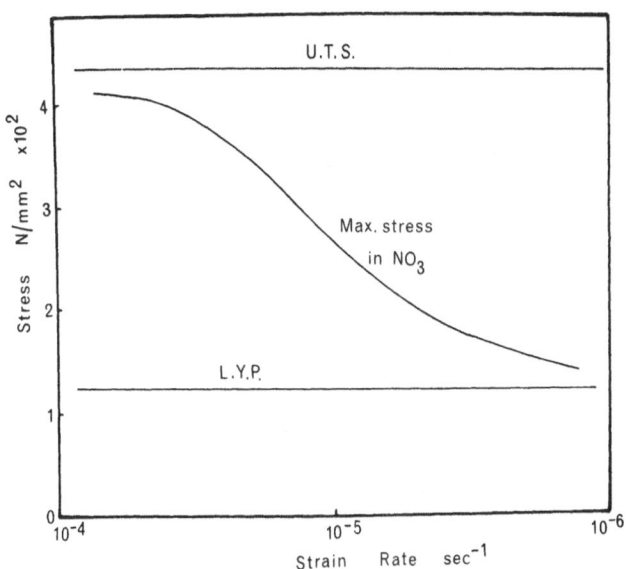

FIGURE 6.   Effects of different strain rates upon
the maximum nominal stress achieved in a
C steel immersed in a boiling nitrate
solution and upon the lower yield point
and ultimate tensile stress in oil at
the same temperature.

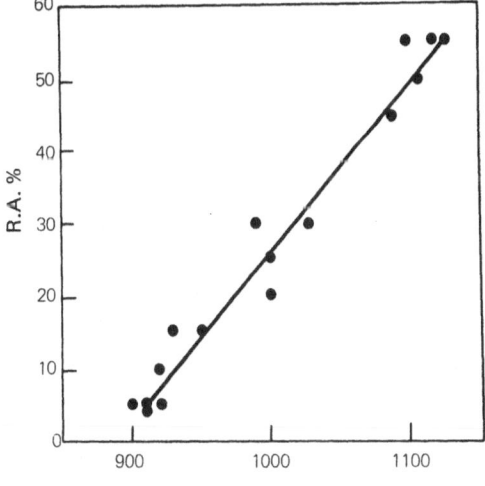

FIGURE 7.   Relationship between the reduction in area
and the time to failure in constant deflection
rate tests upon a C steel immersed in a boiling
nitrate solution.

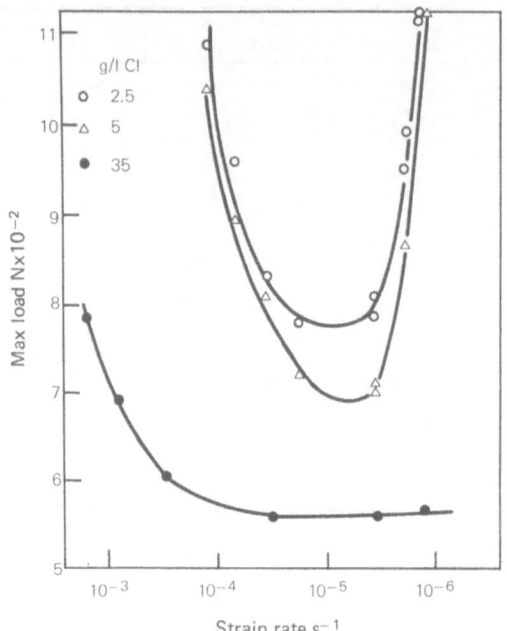

FIGURE 8.   Effects of various strain rates upon the
cracking response of a Mg-Al alloy in
solutions containing 20 gm/litre of chromate
and various amounts of chloride.

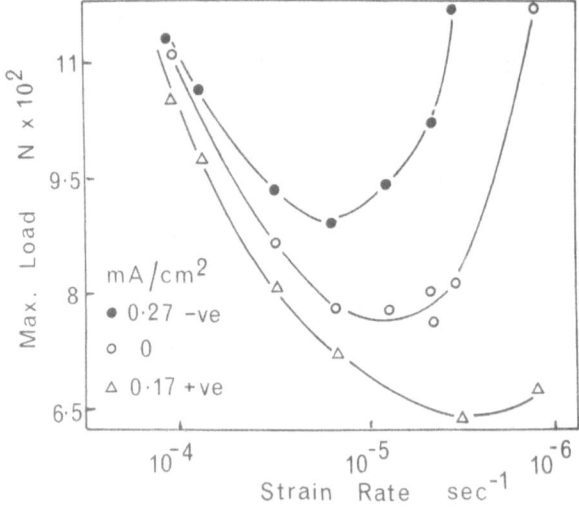

FIGURE 9.   Effects of various strain rates upon the
cracking response of a Mg-Al alloy in a
chromate-chloride solution at various
applied current densities.

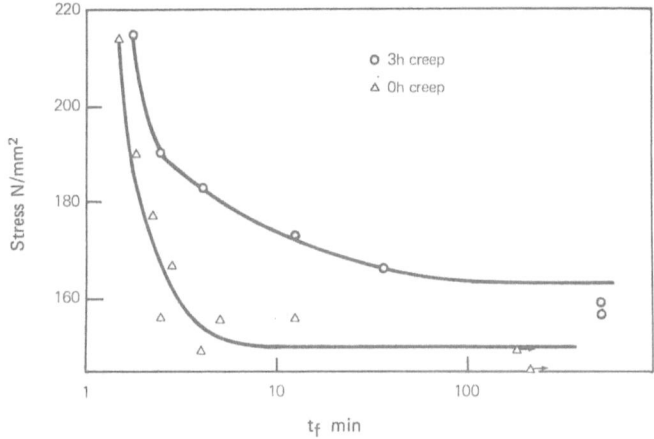

FIGURE 10.   Effect of prior creep in constant load tests
upon the cracking response of a Mg-Al alloy
immersed in a chromate-chloride solution.

FIGURE 11.   Testing machine for constant deflection rate
cantilever bend tests.

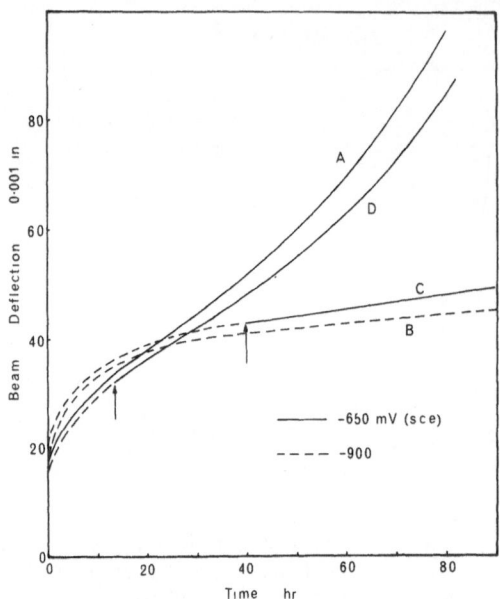

FIGURE 12.   Beam deflection curves from constant load
             cantilever beam tests upon a C steel in
             1N Na$_2$CO$_3$ + 1N NaHCO$_3$ at 75°C and the effects
             of cracking and non-cracking potentials.

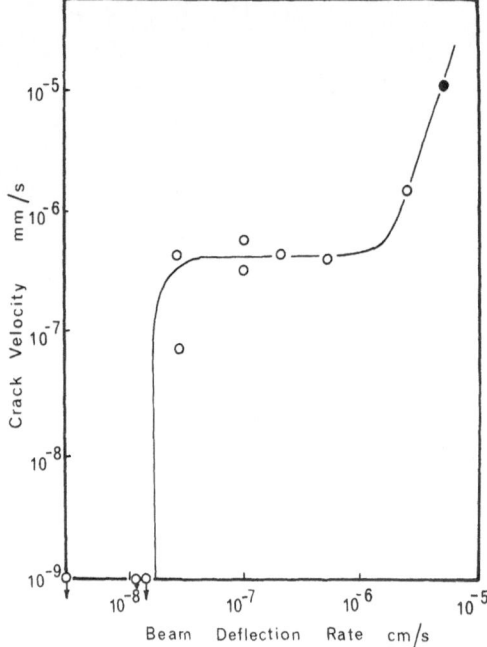

FIGURE 13.   Effects of various applied beam deflection
             rates upon the stress corrosion crack velocity
             for a C steel in 1N Na$_2$CO$_3$ + 1N NaHCO$_3$ at 75°C.

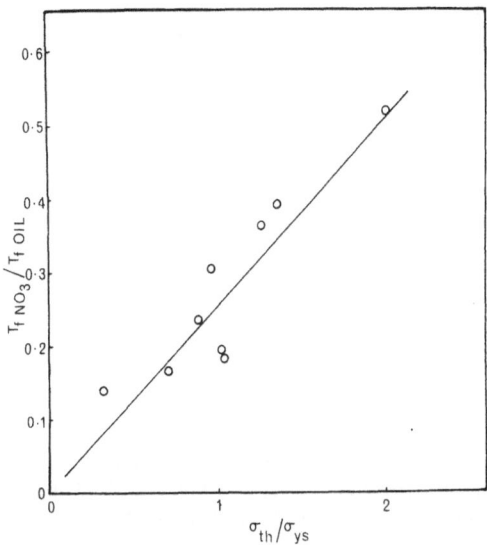

FIGURE 14. Relationship between time to failure ratio
from constant strain rate tests and the
normalized threshold stress from constant
strain tests for a series of low alloy steels
in boiling 4N NH_4NO_3.

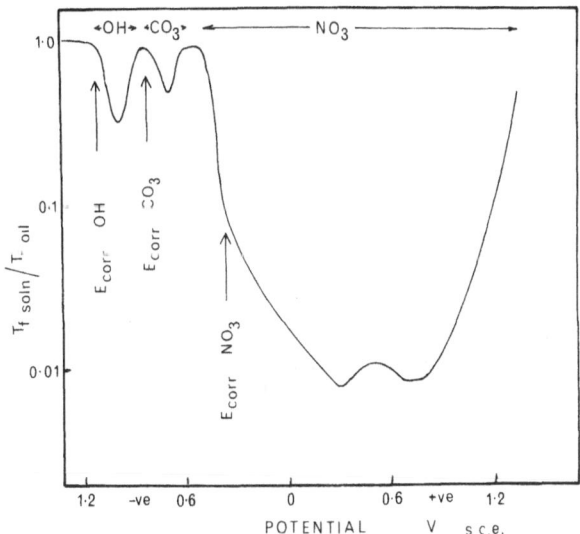

FIGURE 15. Effects of applied potential upon the time to
failure ratio for a C steel in three different
environments.

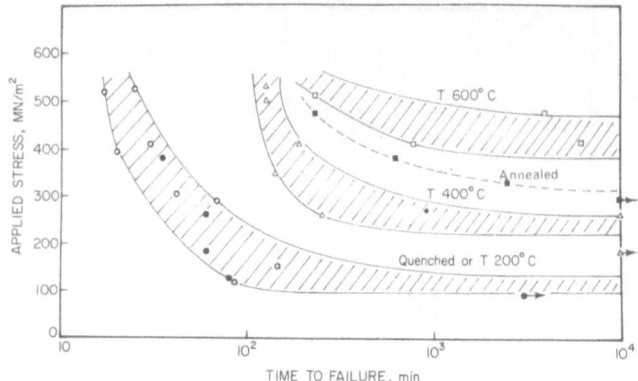

FIGURE 16. Applied stress-time to failure curves in
constant strain tests upon a C steel in a
boiling nitrate solution after various
heat treatments.

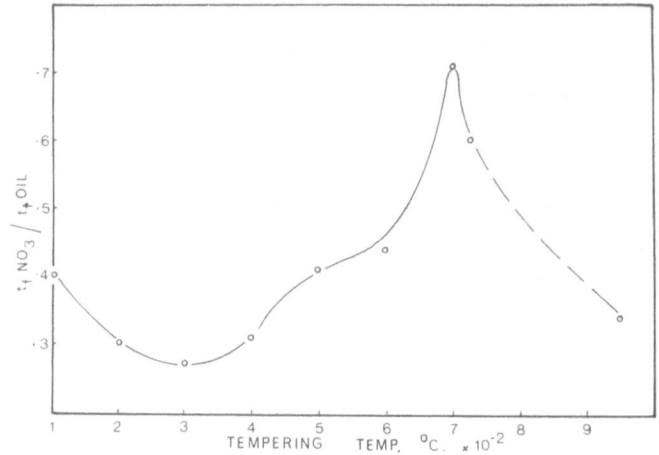

FIGURE 17. Constant strain rate test results upon a
C steel in a boiling nitrate solution after
various heat treatments for comparison with
Figure 16.

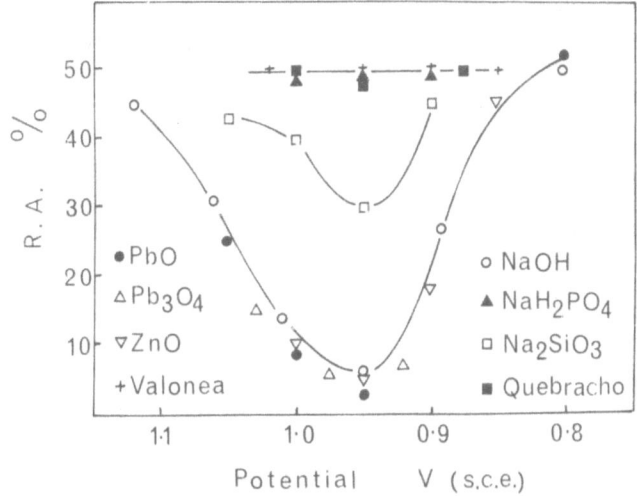

FIGURE 18. Effects of various additions to boiling 35%
NaOH upon the cracking response of a C steel
in constant strain rate tests at different
controlled potentials.

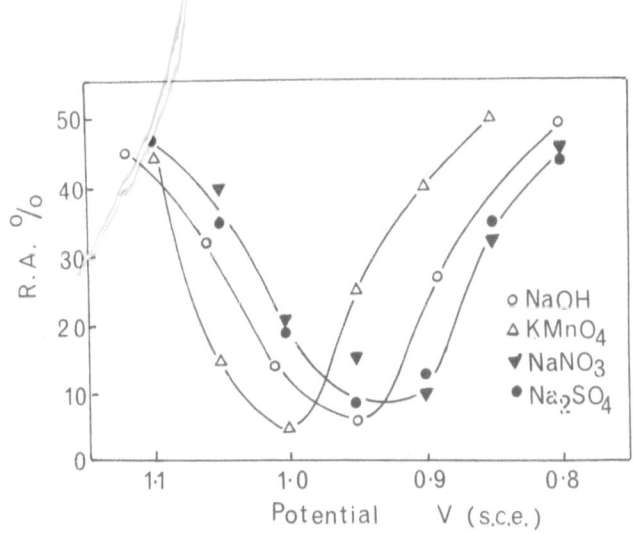

FIGURE 19. Effects of various additions to boiling 35%
NaOH upon the cracking response of a C steel
in constant strain rate tests, showing changes
in the potential range for cracking.

STRESS CORROSION CRACKING OF FERRITIC STEELS

R.N. Parkins

INTRODUCTION

Developments in knowledge and understanding in recent years of the stress corrosion problem in low strength steels has largely conformed to the trends associated with most systems involving environmental sensitive fracture, in that the emphases have been upon the environmental and mechanical aspects of the problem, rather than those concerned with physical metallurgy. The ideas expressed at the time of the NATO Conference in Portugal upon the then emerging working hypothesis[1] that defined the environmental conditions likely to promote stress corrosion cracking have been markedly strengthened by data from different systems, whilst the role of stress, with the ductile low strength steels, is beginning to emerge as related to its effects in promoting slow, but dynamic, plastic strain. The effects of alloying additions to steels, and of some structural variations produced by heat treatments, have received some attention recently and are indicative of the complex interactions that combine to promote stress corrosion cracking.

ENVIRONMENTAL FACTORS

Whilst a review[2] of the environmental aspects of stress corrosion
cracking of low strength ferritic steels indicates that the number of environments that promote cracking continues to increase with time, the concept
of solution specificity remains valid in the sense that some media will
promote cracking but many others do not in a particular steel.   In very
general terms, it is clear that potent solutions will need to promote a
critical balance between activity and passivity, since a highly active
condition will result in general corrosion or pitting, whilst a completely
passive condition cannot, by definition, lead to stress corrosion.   The
borderline conditions between activity and passivity will be most likely
to be achieved for the inherently more corrodable steels by introducing
the passivating tendency through the environment, hence carbon steels
can be made to crack in solutions of anodic inhibitors such as hydroxides
and carbonates.   But for those alloy steels that show a greater degree
of corrosion resistance resulting from the presence of alloying elements,
cracking is likely to be associated with more aggressive environments,
hence the failure of low alloy steels in chloride solutions.   These
combinations of different steels and environments might be expected to
show similar characteristics in terms of their polarization behaviour in
that in each case the range of potentials within which cracking is observed
should be associated with regions of potential in which active → passive
transitions are observed.

There are a number of experimental methods that may be used for
locating regions of active → passive behaviour, but potentiodynamic polarization curves determined at different rates of potential change are
particularly useful in this respect.   By starting with a film free electrode
and sweeping the potential in the anodic direction at a relatively fast rate
($\sim$1000 mV/min), an indication will be obtained of regions wherein intense

anodic activity is likely since there is little time for film growth. If the experiment is now repeated but with, for otherwise identical conditions, a slow rate of potential change ($\sim$10 mV/min), regions will be indicated wherein relative inactivity is likely and a comparison of the two curves will show any ranges of potential within which active surfaces become less so given the time for film formation. Figure 1 shows the potentio-dynamic polarization curves at fast and slow sweep rates for a carbon steel immersed in a carbonate-bicarbonate solution, together with the domains of electrochemical behaviour in different ranges of potential expected from the displacements of the curves on the current scale. At potentials more negative than about -900 mV cathodic protection is implied by the negative currents passed, but above that potential, up to -720 mV, the fast and slow sweep rate curves are not markedly different and there is no suggestion of a reduction of anodic activity with time, implying that pitting or general corrosion will be experienced. Between -720 and -620 mV, the fast sweep rate curve indicates that a bare or thinly filmed surface will be quite active but, given the opportunity in terms of time, the anodic current decays to relatively small values as film formation proceeds and these are the conditions that may be expected to lead to stress corrosion, if other requirements are met, such as that the activity is structurally sensitive and the stress produces an appropriate response in the metal. Above -620 mV, the metal is relatively inactive even at very fast sweep rates, implying the formation of protective films at rates associated with anodic protection. This pattern of expected behaviour is borne out by experiment[3], the potential range in which stress corrosion cracking is actually observed being accurately predicted from the polarization curves.

The designation of potential ranges within which stress corrosion

of ferritic steels is likely, and hence the recognition of potent environments, was originally suggested[4] in the context of the classical cracking
media for C steels, namely nitrates and hydroxides. Subsequently it has
been shown to apply equally to the cracking of Ni containing ferritic
steels in boiling $MgCl_2$[5] and the technique has been used in identifying
CO_3/HCO_3 solutions as cracking media for C steels[3] and acetates as
potent ions in the cracking of a ferritic Cr-Mo steel[6]. Figure 2
shows the fast and slow sweep rate curves for the latter material in
molar ammonium acetate at 90°C and pH8 and the observed cracking range
clearly agrees reasonably with what would be predicted from the polarization curves.

Some idea of the effectiveness of fast and slow sweep rate
potentiodynamic polarization curves in defining the stress corrosion
cracking tendency can be gained from Figure 3. This shows a plot of
the cracking tendency, expressed as a time to failure ratio from slow
strain rate tests, against potential together with a plot of the current
difference between the fast and slow sweep rate curves at various potentials, for the same steel tested in a nitrate, a hydroxide and a carbonate-
bicarbonate solution. Clearly the observed cracking ranges in these
environments are reasonably accurately defined by the polarization
measurements, and furthermore, the severity of cracking, indicated by
increasing departure from a time to failure ratio of 1, reflects the
magnitude of the current density differences in the different solutions.

The active → passive transitions associated with the potential
ranges for cracking can, in some systems, be related to the formation of
soluble ions such as $HFeO_2^-$ in hydroxide solutions and Fe^{++} in carbonate-
bicarbonate solutions and to filming reactions related to the formation
of insoluble substances such as Fe_3O_4 and $FeCO_3$; the most convenient
way of expressing such relationships is in the form of potential-pH

diagrams.   The cracking of C steels in carbonate-bicarbonate solutions

has been studied[7] in sufficient detail to use this approach.   These

solutions vary in pH according to the amounts of carbonate and bicarbonate

ions present but for a given composition the pH is effectively buffered

and remains essentially constant during a stress corrosion test.   Figure 4

shows the experimentally determined potential ranges for cracking in a

number of different carbonate-bicarbonate solutions of various pH, the

cracking domain being defined by two converging lines which predict

that a 1N Na_2CO_3 solution, i.e. without $NaHCO_3$, should not promote stress

corrosion cracking, a prediction that is in line with experimental obser-

vations.   Figure 5 shows a replot of Figure 4 without the experimental

points, but including data from potentiodynamic polarization curves.

The points shown in relation to the upper line indicate the potentials

at which the current fell to relatively low values on fast sweep rate

curves, while the points shown in relation to the lower line indicate

the potentials at which rapid passivation is observed on slow sweep

rate curves.   (The inset on Figure 5 shows the positions on the polariza-

tion curve from which the points were taken.)   The agreement between the

observed cracking domain and that expected from polarization curves is

very reasonable.   Also shown on Figure 5 are the solid phases detected

by electron diffraction from steel surfaces exposed to carbonate-bicarbonate

solutions in appropriate potential ranges.   At more negative potentials

than those delineating the cracking domain, $FeCO_3$ is formed, but is well

known not to provide a protective film, in agreement with the observations

that pitting occurs in its presence and current decay from the initially

active value is relatively small.   At less negative potentials than

those defining the cracking domain Fe_3O_4 is formed and provides anodic

protection.   Within the cracking range both Fe_3O_4 and $FeCO_3$ can form

and since the polarization curves show that Fe^{++} also can form in this range it would appear that the reactions resulting in cracking involve metal dissolution to Fe^{++}, with a critical balance between the relatively high activity that would be associated with $FeCO_3$ alone and the inactivity associated with Fe_3O_4 being provided by a mixture of these substances that changes in proportions in moving from one end to the other of the potential range for cracking. Moreover, the slopes of the lines delineating the cracking domain agree reasonably with the calculated slopes for lines involving interaction between the various substances.

Whilst such electrochemical measurements offer hope of approximately defining likely cracking media and the potential ranges in which these may be expected to produce cracking, there is an element of subjectivity in their interpretation in being non-quantitative and they also suffer from the fact that the observations at any potential will be partly related to the condition of the electrode surface due to the earlier potentials that it has experienced. The latter problem may be surmounted in part by measurements of current decay at specific potentials established rapidly from some initial potential that maintains the surface in a film free condition. When the bare metal is exposed at appropriate potentials the anodic current will rapidly build-up followed by current decay as film thickening occurs. If current decay is very rapid cracking is unlikely, since passivation will be too rapid, whilst if the current decay rate is slow pitting is more likely than cracking because of the sustained anodic activity. An intermediate rate of current decay is likely to be associated with cracking and this is found to be so, but there is as yet no quantitative way of determining independently the rate of decay that will promote cracking and the technique therefore provides no significant advantages over that given by potentiodynamic polarization curves. In systems where initial, air formed, oxide films cannot be reductively dissolved, bare metal has to be exposed either by straining the metal to break the oxide film or by scraping the surface to achieve the same result. Both methods have been used and with some success, although the scraping electrode technique appears more satis-factory than that involved in straining the electrode in that it is likely to provide a more accurately defined area of bare metal. However, both techniques suffer from some of the same advantages as those mentioned earlier in that the contributions to the current response of dissolution and film repair are not readily separable.

The foregoing implies that crack extension is effectively controlled by dissolution at the crack tip, in which case the crack velocity should be related through Faraday's Law to the current densities passed at a bare metal surface. The predicted crack propagation rate, V, is

$$V = i_a \frac{M}{zFd} \qquad \qquad \dots \qquad (1)$$

where i_a is the anodic current density, M = atomic weight, z = valency, F = Faraday Constant and d = density. Taking the effective current density as the largest differences between fast and slow sweep rate polarization curves in the cracking range, Figure 6 shows a plot of these current density differences against observed crack velocities for some ferritic steels in various cracking media the line shown being that calculated from Equation (1), assuming that divalent ions pass into solution. Clearly for a calculation of this type the agreement between observed and calculated crack velocities is very reasonable, especially since the dissolution model involves other facets such as a structural dependence of the attack and the restriction of dissolution to the crack tip by the lateral growth of films over the remainder of the exposed surfaces; this may account for the experimental points of

Figure 6 falling consistently under the calculated line. A further implication of the stress corrosion crack velocity being dependent upon dissolution is that the velocity should have a temperature dependence similar to that for the temperature dependence of the anodic current densities observed in polarization measurements. Figure 7 shows the temperature dependence of the crack velocity of a C-Mn steel in a carbonate-bicarbonate solution, indicating an activation energy in the region of 10 Kcal mol^{-1}, whilst also shown in Figure 7 is a similar Arrhenius plot from polarization curves for the same steel and solution which gives an activation energy of 11 Kcal mol^{-1}. Clearly the close agreement between these activation energies lends strong support to the suggestion that the controlling process in stress corrosion cracking of ferritic steel is crack tip dissolution.

Whilst the dissolution controlled model finds good support from the correspondence between observed characteristics of cracking environments and what would be predicted from thermodynamic and kinetic considerations in relation to the cracking of the low strength, ductile, steels, what is the position in relation to the higher strength ferritic materials? Here, the specificity of cracking environments is much less marked, failure being produced by a wide range of aqueous solutions, organic media and even in gaseous environments that contain hydrogen. The latter is common to all of these environments and the fact that the threshold stress intensities, K_{Iscc}, for cracking in a 3.5% NaCl and in gaseous environments are the same for the Ni-Cr-Mo steel shown in Figure 8 lends strong support to the mechanism involving the embrittlement of the steel in the crack tip region by hydrogen entry. The different times to failure at stress intensities in excess of K_{Iscc} in the different environments to which Figure 8 refers reflects the different plateau crack velocities in these media, being about 0.001, 0.01 and 10 mm/sec respectively in 3.5% NaCl,

in hydrgen gas at 1000 mbar and in hydrogen sulphide gas at 1000 mbar
pressure.   It seems reasonable to expect that species in aqueous
solution which facilitate the ingress of hydrogen into the metal will
enhance cracking, whilst species that lead to the discharge of gaseous
hydrogen at the steel surface will retard cracking.   In the former
category are arsenious salts, whilst platinum additions to the system
may be expected to facilitate hydrogen discharge.   Similarly the effect
of increasing cathodic current densities applied in stress corrosion
tests may be expected to enhance cracking if hydrogen adsorption is
involved in the failure mechanisms.   Figure 9 shows the effects of
sodium arsenate and chloroplatinic acid additions to a NaCl solution
upon the cracking of a 18% Ni maraging steel at various applied cathodic
current densities and clearly indicates conformity with expectations if
hydrogen adsorption is the controlling factor in the cracking process.

The work of Brown[10] involving measurements of pH and potential
at the tips of propagating cracks in a high strength steel, for various
bulk solution pH values and conditions of external polarization, strongly
supports the hydrogen argument.   Thus his work shows that the crack tip
pH is determined by the electrochemical reactions occurring within the
confines of the crack, rather than the bulk solution pH and moreover the
crack tip potential, irrespective of the conditions of external polarization, invariably falls below the line, upon the potential-pH diagram,
that defines the limit of thermodynamic stability of water.   Thus the
conditions at the crack tips were conducive to hydrogen entry into the
steel and control of the pH or potential so that the crack tip conditions
remain above the line relating to the decomposition of water resulted in
no crack growth.   Whilst discussion continues on the details of hydrogen
generation, adsorption and diffusion, and the relative contributions of
these to the overall physical mechanism of hydrogen embrittlement, some

aspects of the latter have begun to crystallize, particularly as the result of experiments conducted in gaseous hydrogen environments. The demonstration that sub-atmospheric pressures of hydrogen gas can readily result in the propagation of cracks in high strength steels indicates that the mechanism is not likely to involve the diffusion of hydrogen through the metal to voids where a disruptive pressure of gas is generated. Of the remaining alternatives, that hydrogen lowers the surface energy by adsorption or that it accumulates within a few atomic distances from the crack tip, in response to the lowering of its chemical potential by the elastic stress, thereby lowering the cohesive force of the lattice, Oriani[11] prefers the latter because of its consistency with the observed effects of small changes in gas pressure and the substitution of deuterium for hydrogen. A sufficient reduction in the hydrogen gas pressure surrounding a high strength steel specimen containing a propagating crack at a given stress intensity will cause the crack to stop propagating, but a subsequent increase in pressure, of about 16 mbar from 217 mbar, was sufficient to restart the crack and with a delay time so short that the extra hydrogen entering the lattice as the result of the increased pressure could have diffused no more than a few atom spacings. Similarly rapid responses of the crack velocity to small changes in applied cathodic current to high strength steels immersed in NaCl solution have been observed. The effect of deuterium in reducing the response to embrittlement appears not to be related to the difference in transport kinetics of the two isotopes but of their solubilities in the dilated lattice just beyond the crack tip[11]. This again is in agreement with a decohesion model, as opposed to one involving surface energy lowering and further support derives from the observation[12] that a stress field having a dilatant component, produced by mode I tensile loading, leads to hydrogen induced cracking, whereas at even higher stress intensities, mode III loading (antiplane shear mode), which has zero dilatant component, does not lead to hydrogen induced cracking.

Whilst these various observations appear to lend strong support to a decohesion model of hydrogen embrittlement, the latter is by no means universally accepted and even if it is there remain points of detail to be elucidated. Thus the observation[10] of the development of acid conditions at crack tips, whilst being indicative of conditions favourable to hydrogen entry into the steel, raises questions as to the extent of dissolution processes in the crack tip region, which also would be facili- tated by the low pH, and the mechanism whereby lateral spreading of the crack is restricted. The association of the potential range for cracking with active → passive transitions for the dissolution controlled model relevant to the cracking of the low strength steels is indicative of filming preventing lateral spread from the crack sides. The evidence available[2] suggests that the films are usually oxides, but the model simply requires a filming substance that will render the crack sides relatively inactive and, whilst different substances will be varyingly effective in that sense, it appears doubtful if the filming substance needs other highly specific properties. Thus, cracking of mild steels in nitrate environments occurs in the presence of films of both Fe_2O_3 and Fe_3O_4 and of various thicknesses, whilst in the same environment low alloy steels containing elements such as Cr can also be caused to crack and in circumstances where presumably Cr is present in surface films. The indications are therefore that the detailed physical nature of the filming substance is secondary to the primary requirement that it renders the underlying metal relatively inactive.

THE FUNCTION OF STRESS

The function of stress in stress corrosion cracking has long remained enigmatic and even over the last decade when this facet of the subject has received more detailed attention, not least because of the advent of fracture mechanics, it has remained a problem in the sense that it is still not possible, for example, to predict theoretically the value of K_{Iscc} for a given system. Especially for the more ductile steels, the threshold stress determined upon initially plain specimens is markedly dependent upon the environment, as well as upon the composition and structure of the steel, so that in boiling nitrate solutions, for example, the threshold stress for a low carbon steel[2] can be as low as about 15 N/mm^2 in 8N NH_4NO_3 and as high as 200 N/mm^2 in 1N $NaNO_3$. These differences may be due in part to variations in the extent to which these different environments promote intergranular corrosion even in the absence of stress, but they also tend to suggest a complementary relationship between stress and environment in stress corrosion cracking. If the function of stress was simply to maintain the crack open to allow the ingress of fresh solution, then the threshold stress for a given material would not be expected to vary much with the composition of the environment, rather like the situation for the high strength steel shown in Figure 8. However, if crack propagation occurs by dissolution at an active tip, with the crack sides rendered inactive by filming, the maintenance of active conditions may be dependent not only upon the electrochemical conditions but also upon the rate at which metal is exposed at the tip by plastic strain. Thus, it may not be stress, per se, but the strain rate that it produces that is important.

The experiments upon fatigue pre-cracked specimens of a C-Mn steel immersed in a carbonate-bicarbonate solution and referred to in the context of Figure 12 of the paper on strain rate testing[13] indicate precisely what would be expected if the crack tip conditions for cracking involved a balance between film growth and the exposure of metal by straining. There is however a further implication of these effects of strain rate in relation to the results from constant load tests. Since cracks will continue to propagate only if their rate of advancement is sufficient to maintain (by the associated stress increase) the crack-tip strain rate above the minimum rate for cracking, it is to be expected that cracks will sometimes stop propagating, and such may particularly occur below the threshold stress, i.e. at stresses below which total failure does not occur. Following tests[14] upon plain specimens of a mild steel immersed in 1N Na_2CO_3 + 1N $NaHCO_3$ and loaded in tension, specimens loaded below the threshold stress which had not failed in an extended period of time were examined metallographically for evidence of cracks, and such were found. Their maximum depth was measured by successive sectioning and the results are shown in Figure 10. In the annealed condition this steel exhibited a threshold stress of 310 N/mm^2, but at stresses below this - down to about the yield stress (240 N/mm^2), cracks of varying depth were detected, with the depth diminishing with initial applied stress. The crack velocities for these deepest cracks were computed from the crack length and the times for which the tests ran, and the results are shown near the points on Figure 10. These crack velocities indicate, by comparison with results mentioned below, that all but possibly the longest crack had ceased to propagate before the completion of the test. It follows that the threshold stress determined in constant-load or constant-strain tests is not the stress below which cracking does not occur, but is that stress at which creep exhaustion following initial loading is faster than the increasing creep resulting from crack extension; thus, the crack tip becomes sufficiently filmed to be rendered inactive, and crack growth ceases. The latter suggestion is supported by the fact

that cracks which have ceased to propagate can be made to propagate again
by a small load increment, which causes plastic strain in the crack-tip
region and reinitiates creep.   Figure 10, in addition to showing the
results already commented upon, shows the effect of surface decarburizing
the same steel in moist hydrogen before stress corrosion testing.   This
treatment reduces not only the threshold stress but also the minimum stress
at which nonpropagating cracks are observed.   While the results in Figure
10 refer to tests in a carbonate-bicarbonate solution, similar results are
obtained in other cracking environments, such as nitrates.

While all of the above observations are in line with what would
be predicted from a model involving the crack tip strain rate as an
important parameter in the mechanism of stress corrosion cracking, the
most convincing demonstration is that in which the cracking parameter
is shown to relate to superimposed strain rate, as opposed to the variable
strain rate that obtains in constant load tests.   Figure 13 of the paper
on strain rate testing[13] refers to such tests, whilst Figure 11 below
shows that the limiting beam deflection rate below which cracking is not
observed is markedly dependent upon the electrochemical conditions.
Since the variations in net section stress during and between these
tests was never by more than a few per cent the effects observed are
unlikely to be related to stress, per se.   But how does the concept of
crack tip strain rate relate to the observation that in boiling 8N NH_4NO_3
the threshold stress for plain specimens is as low as 15 N/mm^2, for a steel
with a yield stress in the region of 200 N/mm^2?   Whilst the yield stress
does not define the lower stress limit for creep at rates of the order of
10^{-7} mm/sec., nevertheless creep at this rate upon loading at 15 N/mm^2
is not likely to be sustained for any significant period of time and it
would appear that threshold stresses in the region of 15 N/mm^2 can only
be rationalized in terms of stress-free intergranular corrosion playing

a large role in the grain boundary penetration process that eventually leads to failure. Such intergranular corrosion is known to occur in mild steels exposed to nitrates in appropriate circumstances[15], but in the less potent nitrates and in environments such as those based upon hydroxides or carbonate-bicarbonate mixtures the extent of grain boundary corrosion in the absence of stress is very small before the reactions are stifled[1]. In these latter environments stress corrosion cracking can only occur if the initial applied stress exceeds the yield stress and in some cases cannot be produced in tests at constant strain or constant load but only when a constant deflection rate is applied to the specimens[13]. Moreover, in nitrates of intermediate potency, the grain boundary penetration rate increases sharply when the penetration reaches the depth where the yield stress is achieved in the remaining specimen section i.e. when an open crack is produced by yawning[16]. These various observations are all consistent with a model involving complementary roles for plastic deformation and the electrochemical reactions at the crack tip. However, the quantification of the model presents difficulties at present, particularly in defining the relevant electrochemical conditions at the crack tip. If a precipitated film forms over the crack tip and this is repetitively ruptured by plastic strain in the underlying metal, it is possible to derive an expression[17] with the time interval between rupture events playing a critical role in relation to the crack propagation rate. But results such as those shown in Figure 6 suggest that the crack propagation rate is essentially controlled by the dissolution rate, rather than by the rate at which precipitated films form, although these two rates are not unconnected. Moreover, the rate at which current decay occurs on initially bare surfaces is such that the surface becomes relatively inactive long before precipitated films are visible. It appears likely therefore that the crack tip conditions should be represented by a viscous layer that is thinned, but not broken, by strain in the crack tip region, the thinning process facilitating further dissolution.

The situation in relation to the higher strength steels is more promising. Thus, Gerberich and Chen[18] have derived an expression for K_{Iscc} in terms of the ratio of the critical hydrogen content for embrittlement in the crack tip region, C_{cr}, to the initial uniform concentration of hydrogen, C_o, of the form

$$K_{Iscc} = \frac{RT}{\alpha V_H} \ln \left(\frac{C_{cr}}{C_o} \right) - \frac{\sigma_{ys}}{2\alpha} \qquad \dots \qquad (2)$$

where α is a constant related to the plastic constrain factor, V_H is the partial molal volume and the other symbols have their usual significance. This shows good correlation with experimental results for, for example, a range of yield strengths in 0.4 C steels, as shown in Figure 12. The trend of such results, showing increased susceptibility to cracking with increasing yield stress, may be interpreted as indicating that cracking is more likely as the tendency towards plastic deformation is decreased, but there is evidence[19] to show that cracking is associated with the formation of stretch zones in the region of the pre-crack tip. Moreover, van Leeuwen[20], in a quantitative treatment of the plateau crack velocity in the failure of high strength steels, regards the occurrence of a plateau velocity as resulting from the development of a plastic enclave at the crack tip and to the associated crack blunting, as well as to the limited diffusion rate of hydrogen and the dependence of the critical concentration of the latter for cracking upon the stress intensity.

EFFECTS OF STEEL STRUCTURE AND COMPOSITION

While transgranular stress corrosion cracking of carbon and low
alloy steels does occur in some environments, e.g. $MgCl_2$ in the case of
ferritic Ni steels and Na_2CO_3-$NaHCO_3$ solutions in the case of Ti containing
steels, the more common mode of failure is intergranular.   Since the
crack paths coincide with paths susceptible to corrosion in the same
environmental conditions but in the absence of stress, the implication
is that low strength steels contain pre-existing paths which are activated by the application of stress.   The most obvious reason for
structurally dependent corrosion is the presence of electrochemically
distinct segregates or precipitates at the grain boundaries, and there
is much evidence in relation to annealed or quenched and tempered steels
that cracking propensity is related to the presence, or otherwise, of
soluble carbon or precipitated carbides at the ferrite grain boundaries[1].
This effect of carbon content upon cracking propensity can be explained
in terms of its distribution in the steel.   Thus, in low C steels
($< \sim 0.1\%$ C) the carbon segregates to the ferrite grain boundaries either
as soluble carbon or as carbide particles or films, if the solubility
limit of carbon in ferrite is exceeded.   As the carbon content is
increased, so the proportion of pearlite in the steel increases and the
number of carbide particles in the boundary decreases.   It may be expected,
therefore, that heat treatments which alter the distribution of the carbon
in a given steel will alter the stress corrosion propensity, and this is
found to be so.   Subcritically annealing a relatively resistant grade of
steel, with the carbide present as pearlite, results in spheroidisation
and dispersion of the carbide in globular form, particularly to the ferrite
boundaries, as the annealing time is increased;   as the results in Figure 13
indicate, there is an attendant marked increase in the cracking suscepti-

bility of the steel. Similarly, when carbon steels are quenched and tempered at various temperatures, cracking susceptibility appears related in part to the effects of these treatments upon carbide dispersion[21]. Figure 17 of the paper on strain rate testing[13] shows the cracking susceptibility, as measured in slow strain rate tests, for a C steel quenched and tempered for 1 hour at various temperatures. The most resistant condition results from tempering at 700°C, which produces a very fine dispersion of carbide upon the prior austenite boundaries and the prior martensite lath boundaries, with the result that stress corrosion cracks develop multiple fine branches and the macroscopic crack velocity is reduced. However, tempering at higher temperatures, or for longer times, causes the ferrite to recrystallise and the carbide particles to grow and be preferentially located at the ferrite grain boundaries, with an attendant increase in susceptibility to cracking; this is in agreement with the earlier observations concerning the effects of prolonged subcritical annealing of pearlitic structures.

This function of carbon has, not surprisingly, resulted in suggestions that the role of further alloying additions to steel is related to the carbide forming or graphitizing tendencies of these further additions. Thus, Long and Lockington[22] suggest that the resistance to cracking in nitrates resulting from Cr or Ti additions to a very low C iron and the maintained susceptibility in the presence of Ni or Al are related to the carbide-forming tendencies of the former pair and the graphitizing characteristics of the latter pair of elements. While there may be some relationship between susceptibility and the presence or otherwise of very strong carbide formers, such as Ti, the effects of this and other alloying elements may relate to other factors. Thus, structure not only relates to the distribution of local cells but also to mechanical properties and the latter relate to the response of the metal to the application of stress.   Similarly, alloying additions

may influence film formation or dissolution kinetics, quite apart from

any effect that they have upon local cell distribution.   Moreover, in

relation to any electrochemical influences of alloying elements, there

is no reason to expect that their effects will be the same in a range

of different stress corrosion cracking environments.

To study fully the effects of alloying additions to steels upon

stress corrosion resistance it is important to realize the limitations

of tests in a single environment at the corrosion potential.   The effects

of potential upon cracking resistance have already been indicated (Figure 3)

and it is conceivable that the presence of a particular alloying element

could cause the free corrosion potential to move outside the cracking

range in one environment and to a more susceptible potential in another

solution, from which it may be deduced that in the former event the

alloying was beneficial and in the latter event detrimental.   However,

since the actual potential achieved in service may be very different from

that obtaining in a laboratory experiment, the above deduction may be

irrelevant.   The most effective way of determining the effects of alloying

additions clearly is to conduct experiments over a range of potential in

different environments.

Figure 14 shows the effects of Al, Cu and Ti additions to a

0.1% C steel upon cracking in boiling 4N $NaNO_3$ at various potentials.

The beneficial effect of Ti in reducing the potential range for cracking

and in reducing the cracking tendency at the most susceptible potential,

as compared with the carbon steel, are readily apparent, as are the facts

that Al has a slight beneficial effect and Cu no effect.   Limited test

results for other alloying additions to a higher carbon steel and for a

carbonate-bicarbonate environment are shown in Figure 15, all of the

alloying conditions showing a beneficial effect in varying degrees.   It should not be assumed from the curves of Figures 14 and 15 that alloying additions, while sometimes not beneficial, are never harmful in the sense that they promote increased susceptibility to cracking since, for example, molybdenum additions markedly increase susceptibility to cracking in hydroxide solutions by extending the potential range for cracking by several hundred millivolts.   Similarly, alloying additions may induce a susceptibility to cracking in an environment that does not promote cracking of carbon steels.   Thus, the latter are not susceptible to cracking in boiling $MgCl_2$ solutions, an environment commonly used in laboratories for assessing the cracking response of austenitic steels, but the addition of only 1% of Ni to a ferritic steel induces a marked susceptibility to cracking in such an environment[5].

In general terms, then, it is apparent that different alloying additions to steels produce different cracking responses and that the magnitude and trend of the latter may vary with the environment involved. It is to be expected that the amount of the alloying additions will be important, also, the results shown in Figure 16 indicate that this is so for Cr additions in relation to cracking in boiling 4N $NaNO_3$.   In Table I, an attempt has been made to assess the effects of different alloying elements added to ferritic steels upon the cracking susceptibility in three different environments.   Inevitably, in view of the method of rating the effects of the different alloying additions there is an element of subjectivity in the results recorded, especially at the borderlines; furthermore, it should be remembered that because of the earlier mentioned effect of carbon as an alloying addition there is the possibility of the elements listed in Table I producing different responses at different carbon

Table I. Effects of alloying additions to ferritic
steel upon stress-corrosion cracking in
various environments.

Element	Environment		
	NO_3	OH	CO_3/HCO_3
Al	3	3	3
Cr	2	2	1
Cu	4	3	3
Mo	4	4	1
Ni	2	1	1
Si	3	4	3
Ti	2	2	1

Rating :   1   2   3   4
         Good Improved No Change Bad
(As compared with given carbon
content).

contents.   Thus, the deleterious effect of copper additions to cracking
in nitrate solutions is much more marked at low ($\sim$0.05%) than at high
($\sim$0.25%) carbon content.   Notwithstanding these difficulties, the
ratings shown in Table I give a broad indication of trends and show that
the most beneficial elements vary with the environment considered and are
not simply related to the carbide-forming tendencies of the elements. Of
course, the strong carbide formers, such as titanium and molybdenum, alter
the structure of ferritic steels when present at appropriate levels by
promoting a globular carbide distribution as opposed to the more usual
laminated pearlite;   this may influence cracking response either by changing
the distribution of local cell action or through influencing the mechanical
properties of the steel, as discussed later.   It would appear most likely,
however, in view of the varying effects of alloying with change in environment, that the role of alloying in relation to dissolution and/or film
formation will be important.   This is supported by the results shown
in Table 2, which indicate the maximum anodic current densities at

relatively bare surfaces and the time-to-failure ratio from constant
strain rate tests in boiling 35% NaOH for steels containing Mo, Cr, or
Ni, together with the results for a carbon steel for comparative purposes.

Table 2. Effects of alloying additions to ferritic
steel upon maximum anodic current density
and time to failure ratio for cracking in
boiling 35% NaOH.

Steel	Steel			
	0.1% C	5% Mo	1.75% Cr	6.05% Ni
Maximum Current Density $\mu A/cm^2$	62	50	40	11
$T_f OH/T_f$ Oil	0.22	0.36	0.66	0.90

Clearly, the particularly beneficial effect of Ni additions in relation
to cracking in hydroxide solutions is reflected in these results, as
indeed are the less marked trends resulting from other additions. Fast
and slow sweep rate polarization curves also indicate correlation between
electrochemical parameters and cracking propensity, where the latter is
significantly changed by the alloying addition. Thus, fast and slow
sweep rate curves for a steel containing 5% Mo immersed in boiling 35%
NaOH, in which the range of potentials for stress corrosion cracking is
considerably extended by comparison with a carbon steel, show by comparison
with the corresponding curves for a carbon steel, that in the same environment the Mo addition produces a second active peak at less negative potentials
and that the current densities passed at a high sweep rate remain relatively
high, indicating that cracking is likely over a potential range of about
1 volt, as is indeed observed.

While it appears that the electrochemical changes resulting
from alloying ferritic steels will be of considerable importance in
determining cracking response, the effects upon mechanical properties
of the addition of alloying elements may also be expected to be signifi-

cant, particularly so when the amount or nature of the alloying element produces only small electrochemical changes. This interdependence of electrochemistry and response to the application of stress in low alloy steels may be illustrated by the crack velocity-beam deflection rate curves of Figure 17 for strain rate tests upon precracked specimens immersed in 1N Na_2CO_3 + 1N $NaHCO_3$. Clearly, 1% Cr or 2% Ni added to a 0.25% C steel lowers the plateau crack velocity quite appreciably - by about an order of magnitude in the case of the Ni addition, while the limiting beam deflection rate is also changed significantly; both parameters reflect the higher stress corrosion crack resistance resulting from the presence of these alloying elements. Figure 16 also shows that the 1.4% Ti steel did not sustain stress corrosion cracking in these tests and that such an addition produces a very resistant steel so far as cracking in carbonate-bicarbonate solution is concerned. Perhaps the most direct illustration of the significance of alloying upon mechanical properties in relation to stress corrosion cracking is shown by the results indicated in Figure 18 for some steels tested in boiling 4N $NaNO_3$ solution. The threshold stress, determined in constant strain tests, has been divided by the yield strength for a series of ferritic steels containing different amounts of various alloying additions, and the fact that so many of the steels show the ratio of these stresses to be close to 1 indicates the importance of plastic yielding in determining the threshold stress. At sufficiently high concentrations of a beneficial element such as Ti or a detrimental element such as Cu, the ratio of the stresses deviates from 1, because the electrochemical influences of these additions becomes more dominant at the higher concentrations. As a final illustration of the importance of mechanical properties in determining cracking response, the effect of grain size upon cracking in ferritic steels may be cited[15]. Figure 19 shows the effects of grain size upon the lower yield stress and the 5% flow stress for a carbon steel, together with the dependence of stress corrosion cracking in a boiling nitrate solution upon ferritic grain size. The similarities of the slopes of all these stresses with grain size is a strong indication that the greater cracking susceptibility of coarse grained steels is related to their inferior mechanical properties when compared with the same steel in a fine grained condition.

These various effects of steel structure and composition should serve to make the general point that, any attempt to interpret the effect upon cracking response, arising from a change in the steel, in terms of an influence upon a single parameter is likely to be misleading. The same is probably true for the higher strength steels where changes in structure or composition may influence mechanical behaviour, the release of hydrogen from solution or its adsorption, diffusivity or solution in the steel, but these are matters that will be dealt with in other papers.

CONCLUSION

The stress corrosion of ferritic steels is the result of critical balance being achieved between a number of interdependent parameters. The essential requirements for cracking, of localized dissolution or hydrogen interaction at the crack tip coupled with relative inactivity of all other exposed surfaces, imply interactions between the environment and the metal and the response of the latter to the application of stress that under-line the interdependence of the controlling parameters. Thus the active → passive transition that must occur on the crack sides generated by the advancing crack is not only dependent upon the solution composition but also upon the metal composition and the extent to which the metal is exposed by plastic deformation resulting from the stress. This is in turn dependent upon metal structure and composition, to complete the cycle of interdependence of controlling parameters. In principle the

control of any one of these parameters leads to alleviation of the
problem. While metallurgical control of the problem presents an
obvious approach, this will not always be the cheapest solution since,
in relation to alloying, for example, the amount of addition required
to markedly improve cracking resistance will usually be in excess of
about 2%. Influencing the environmental reactions, through control
of solution composition, potential or temperature, or avoiding those
stress conditions that result in cracking, will sometimes be cheaper,
and at least as effective, as alloying as a means of controlling stress
corrosion cracking.

REFERENCES

1. R.N. Parkins, The Theory of Stress Corrosion Cracking in Alloys, NATO, 1971, p.167.

2. R.N. Parkins, Proc. Conf. on "Stress Corrosion Cracking and Hydrogen Embrittlement of Iron Base Alloys". Firminy, France, 1973. To be published by NACE.

3. J.M. Sutcliffe, R.R. Fessler, W.K. Boyd and R.N. Parkins, Corrosion, 28, 1972, 313.

4. M.J. Humphries and R.N. Parkins, Fundamental Aspects of Stress Corrosion Cracking, NACE, 1969, p.384.

5. B.S. Poulson and R.N. Parkins, Corrosion, 29, 414, 1973.

6. J.G. Parker and W.G. Pearce, Corrosion, 30, 18, 1974.

7. G. Marsh and R.N. Parkins, unpublished results.

8. T.P. Hoar and J.R. Galvele, Corr. Sci., 10, 1970, 211.

9. A. Brown, J.T. Harrison and R. Williams, as Reference 2 above.

10. B.F. Brown, as Reference 1 above, p.186.

11. R.A. Oriani, as Reference 2 above.

12. C. St. John and W.W. Gerberich, Met. Trans., 4, 589, 1973.

13. R.N. Parkins, "Constant Strain Rate Stress Corrosion Testing" - this Conference.

14. R.N. Parkins, 5th Symposium on Line Pipe Research, American Gas Association, Catalogue No. L30174, 1974.

15. M. Henthorne and R.N. Parkins, Brit. Corr. Jnl, 2, 186, 1967.

16. M. Henthorne and R.N. Parkins, Corr. Sci., 6, 357, 1966.

17. D.A. Vermilyea, as Reference 2 above.

18. W.W. Gerberich and Y.T. Chen, Metallurgical Transactions A, 6A, 271, 1975.

19. P. McIntyre, as Reference 2 above.

20. H.P. van Leeuwen, Corrosion, 31, 42, 1975.

21. R.N. Parkins, P.W. Slattery, W.R. Middleton and M.J. Humphries, Brit. Corr. Jnl., 8, 117, 1973.

22. L.M. Long and N.A. Lockington, Corr. Sci., 11, 853, 1971.

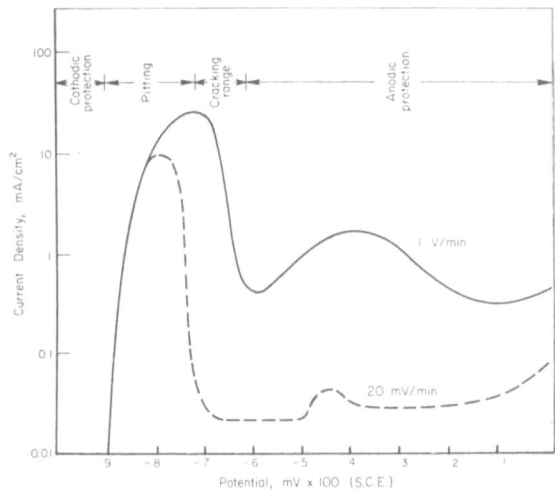

Figure 1.   Potentiodynamic polarization curves for a
            C steel in 1N Na₂CO₃ + 1N NaHCO₃ at 90°C,
            showing domains of electrochemical behaviour.

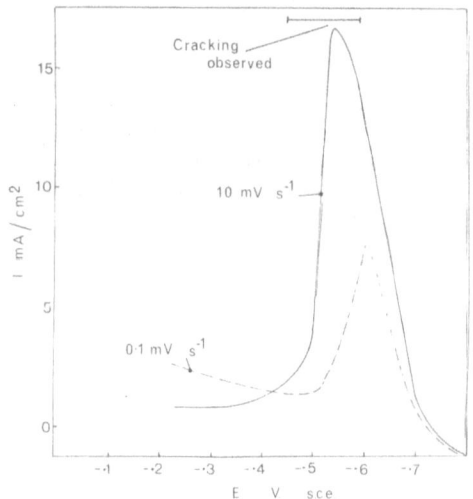

Figure 2.   Potentiodynamic polarization curves for a
            CrMo steel in molar ammonium acetate at
            90°C, pH8, and the observed cracking range[6].

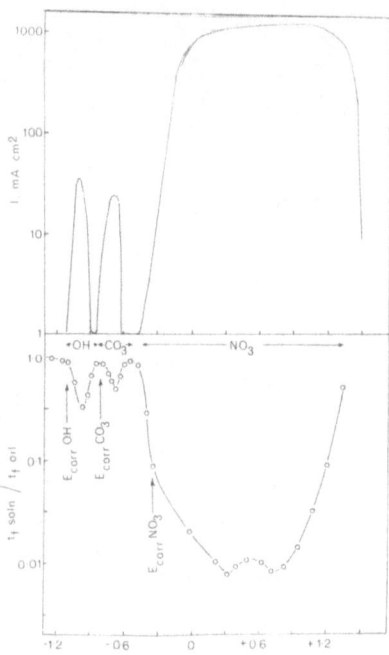

Figure 3.   Showing the correspondence between observed
            cracking ranges and those predicted from
            electrochemical measurements for a C steel in
            boiling 33% NaOH, boiling 20% NH$_4$NO$_3$ and
            1N Na$_2$CO$_3$ + 1N NaHCO$_3$ at 75°C.

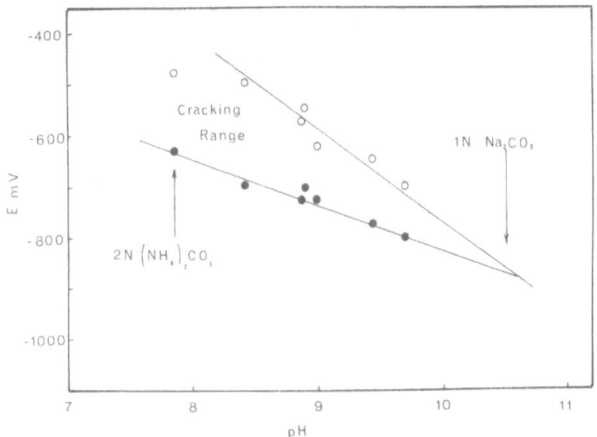

Figure 4. The pH dependence of the intergranular stress corrosion cracking range for a C steel in various CO_3/HCO_3 solutions at 75°C.

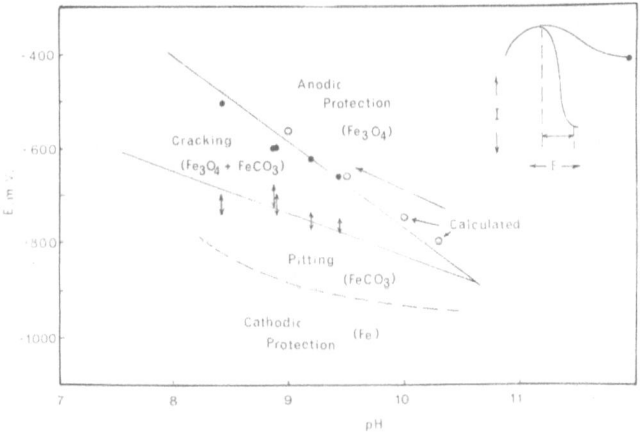

Figure 5. Comparison of the results from stress corrosion tests with those from polarization measurements for various CO_3/HCO_3 solutions, showing the extent to which the experimentally observed cracking range can be predicted from electrochemical measurements.

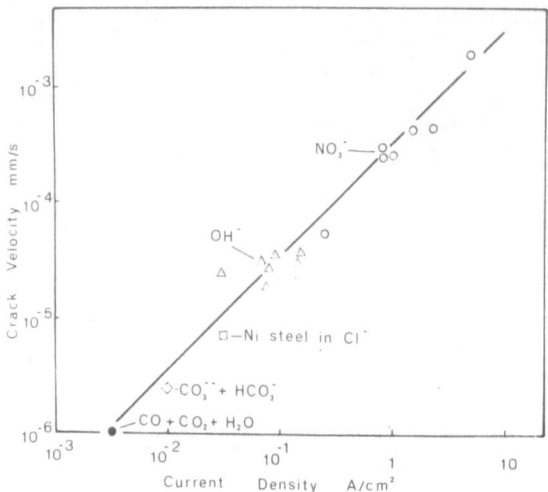

Figure 6.   Experimentally determined crack velocities and maximum current densities for ferritic steels in various environments, including data from various sources(8,9,14). (The line is calculated from Equation (1).)

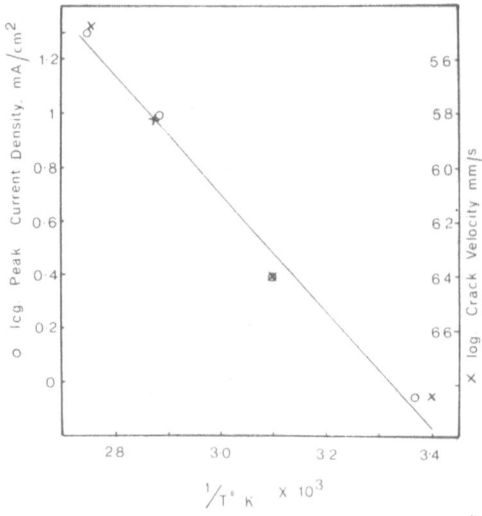

Figure 7.   Temperature dependences of crack velocity and peak current density on bare metal surface for C steel in 1N Na_2CO_3 + 1N $NaHCO_3$ showing similar activation energies.

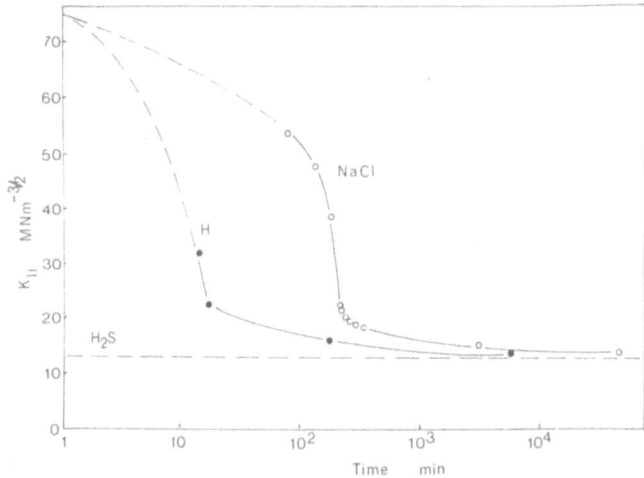

Figure 8.   Time to failure of NiCrMo steel as a function
of initial stress intensity in gaseous H_2S,
gaseous H and 3.5% NaCl solution[19].

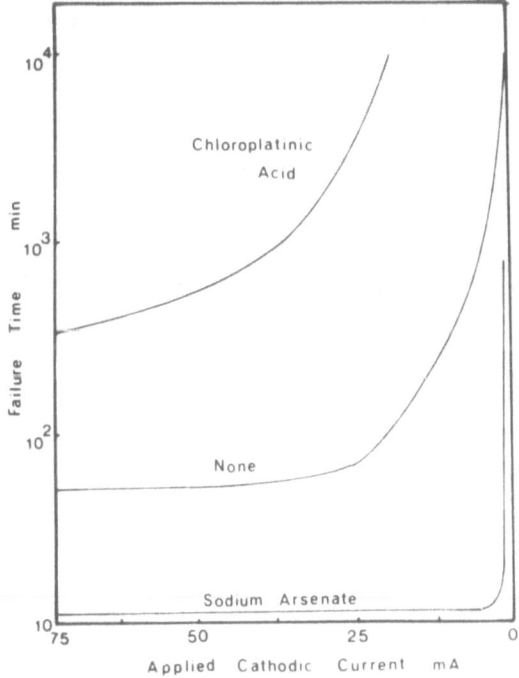

Figure 9.   Effects of additions to 0.6N NaCl, pH2, upon
time to failure of initially plain specimens
of a 18% Ni maraging steel.

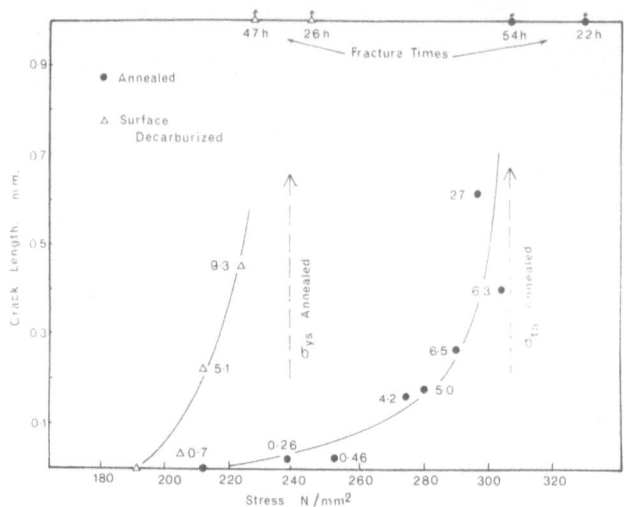

Figure 10.   Maximum stress corrosion crack lengths for
different applied stresses in constant load
tests on annealed and surface decarburized
0.05% C steel in 1N Na_2CO_3 + 1N $NaHCO_3$ at 90°C
and -650 mV (s.c.e.).   (The numbers against the
points are the apparent crack velocities x 10^7
mm/sec.)

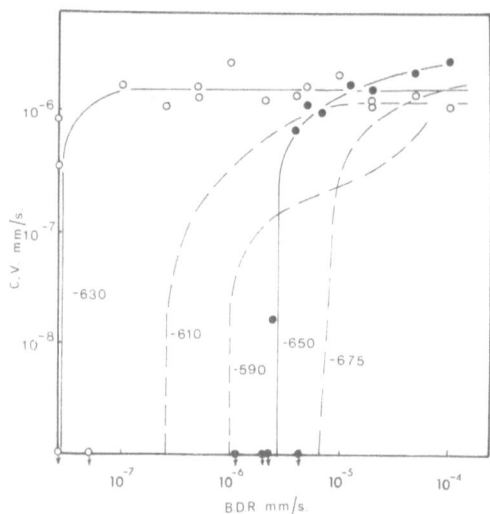

Figure 11.   Effects of different potentials upon the
applied beam deflection rate - crack velocity
curves for a C steel in 1N Na_2CO_3 + 1N $NaHCO_3$
at 75°C.

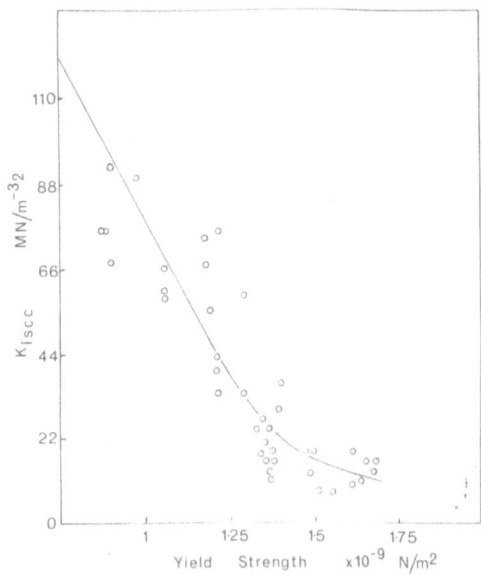

Figure 12. Effect of yield strengths upon K_{Iscc} values
for 0.4 C low alloy steels tested in distilled
water or 3% NaCl solution[18]. (The line is
calculated from Equation (2).)

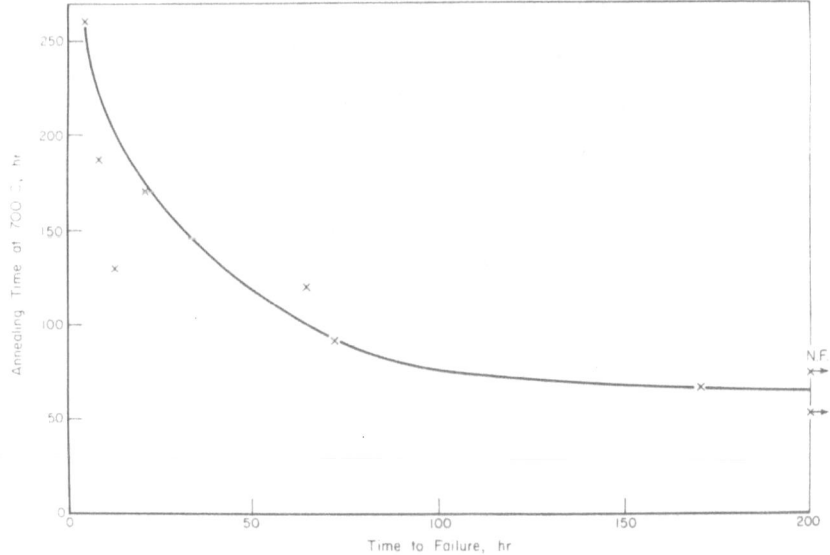

Figure 13. Effect of time of subcritical annealing at
700°C upon the cracking susceptibility of a
0.22% C steel in a boiling nitrate solution.

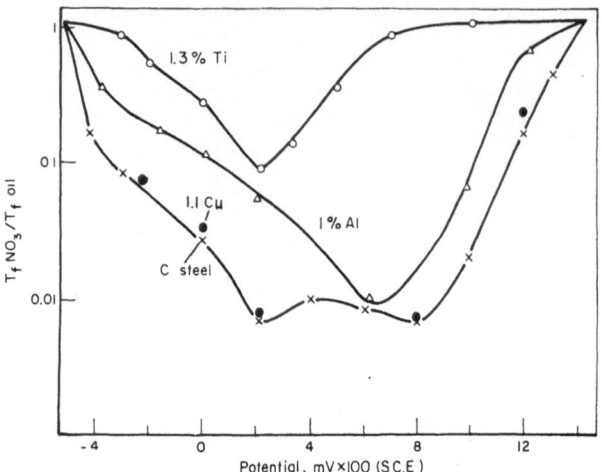

Figure 14.   Effects of Al, Cu and Ti additions to a
             ferritic steel upon the cracking susceptibility
             in boiling 4N NaNO₃.

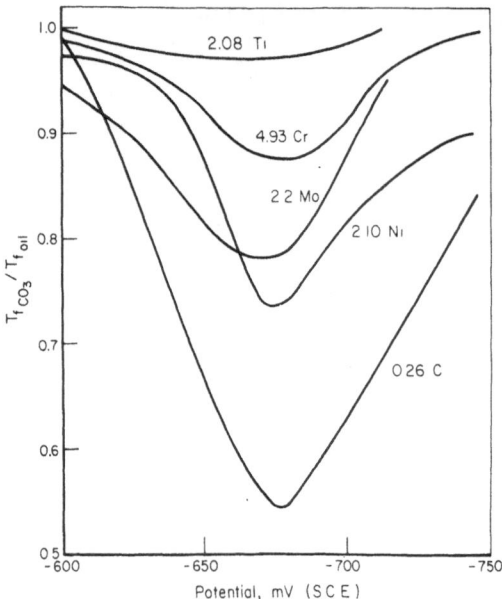

Figure 15.   Effects of various alloying additions to a
             ferritic steel upon the stress corrosion
             cracking susceptibility in 1N Na₂CO₃ +
             1N NaHCO₃ at 75°C.

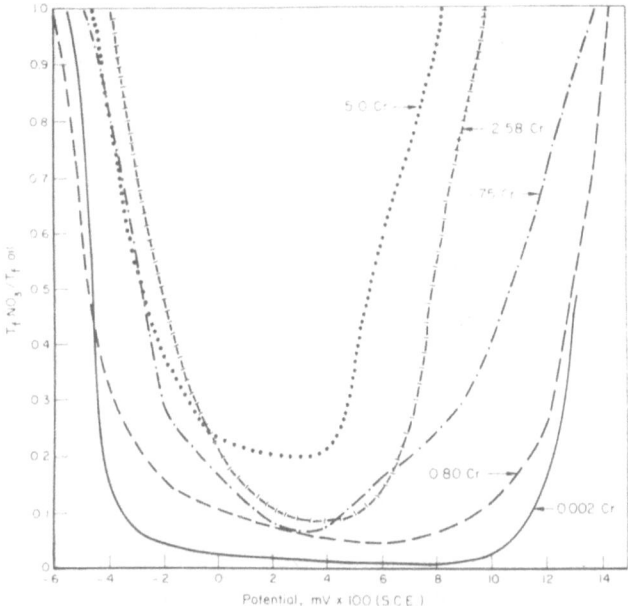

Figure 16.   Effects of different Cr additions to a
ferritic steel upon cracking susceptibility
in boiling 4N NaNO₃.

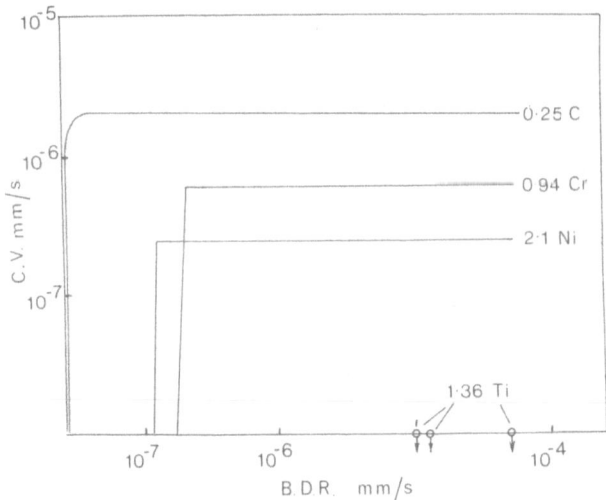

Figure 17.   Effects of various alloying additions to a
ferritic steel upon the applied beam deflection
rate - crack velocity curves while immersed
in 1N Na₂CO₃ + 1N NaHCO₃ at 75°C.

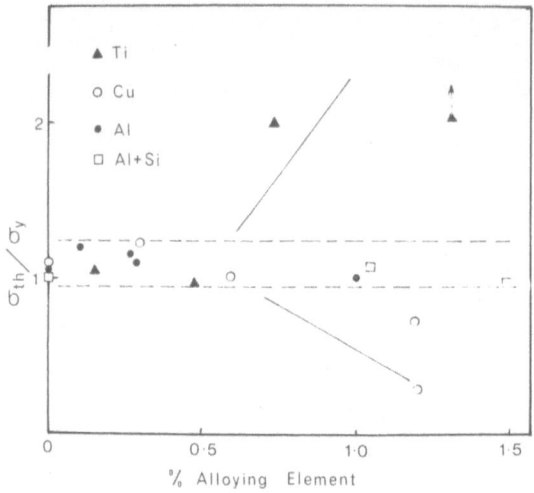

Figure 18. Normalized threshold stresses for cracking of low alloy steels in boiling 4N NaNO₃, as a function of alloy content.

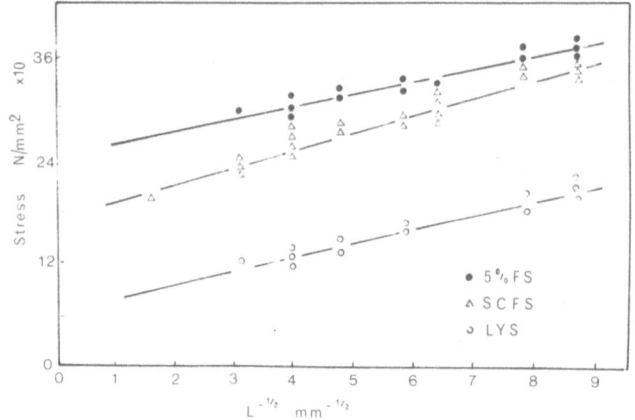

Figure 19. Effects of grain size upon yield and stress corrosion fracture stresses for C steel tested in a boiling nitrate solution[15].

THE FRACTURE MECHANICS APPROACH TO STRESS CORROSION CRACKING

by R. P. Wei

Department of Mechanical Engineering and Mechanics
LEHIGH UNIVERSITY
Bethlehem, Pennsylvania 18015 USA

ABSTRACT

The fracture mechanics approach to stress corrosion cracking
has evolved over the past ten years. It has now been recognized as
an important tool for structural design and for use in establishing
the fundamental mechanisms for stress corrosion cracking. Standard
test methods and guidelines are now being developed to ensure the
orderly development in this field. In the meantime, it is essential
that the basic assumptions of fracture mechanics analysis and the
various problems associated with stress corrosion testing are pro-
perly understood, so as to ensure proper utilization of test
methods, and correct interpretation and usage of results.

The fracture mechanics based approaches to stress corrosion
cracking and the essential assumptions of linear fracture mechanics
analysis are reviewed. Test methods and specimen design considerations are summarized in light of the analysis assumptions. Specific problems in testing and in data interpretation related to
materials response are reviewed. The implications of these pro-

blems and of the fracture mechanics approach for stress corrosion
cracking studies are discussed.

INTRODUCTION

Fail-safe and safe-life principles form much of the basis for
modern design philosophy, particularly in the aerospace industry,
and are being adopted in many other industries. Both of these
principles recognize that cracks may be present in a structural
component, or may initiate early in its service life, and that these
cracks may grow during service. Structural design, therefore, must
incorporate considerations of both the strength and the durability
of a component. The strength, or the load bearing capacity of a
component in the presence of cracks or crack-like defects, is a
function of the material's fracture toughness[1-4]. The durability
or useful life of the component is governed by the rate of degrada-
tion of the load bearing capacity of such a component as a result
of subcritical crack growth, and is a function of the subcritical
crack growth resistance of the material [5-7]. (Subcritical crack
is defined as a crack that is smaller in size than that required to
cause fracture under the prevailing applied stress - the critical
crack). Subcritical crack growth can occur under both sustained
and cyclically varying loads; more commonly referred to as stress
corrosion cracking and fatigue (or corrosion fatigue). Here, only
the stress corrosion cracking aspect of durability will be considered.

Figure 1 serves to illustrate the respective roles of fracture
toughness and subcritical crack growth resistance in determining
structural integrity and durability. Fracture toughness measures
the tolerance of a material for cracks under load and, as such,
defines the condition for failure or the onset of unstable fracture.

Quantitative definition of fracture toughness has now been established, through the formalism of linear fracture mechanics analyses, in terms of the critical stress intensity factor (K_{Ic} or K_c), or the critical strain energy release rate (G_{Ic} or G_c) [1-3], or more generally, in terms of the critical strain-energy-density factor (S_c) [4]. These parameters can be readily measured in the laboratory (for high strength materials with moderate toughness) and serve to define a failure locus as illustrated in Fig. 1. (Failure by general yielding, or plastic flow instability, will not be considered here).

For a crack of given size, a specific critical stress for the failure of a material is defined by its failure locus (or, fracture toughness); for example, a_1 and σ_1. Conversely, for a given stress, a critical crack size is similarly defined: σ_2 and a_2. The strength (or residual strength) of the structural component is, therefore, defined by the size of the dominant crack that is present and by the fracture toughness of the material. In principle, the region to the left of the failure locus represents the 'safe' region with respect to catastrophic failure. For example, it should be safe to load a structural component containing a crack of size a_1 to some stress level, say σ_2 (where $\sigma_2 < \sigma_1$), since a_1 is smaller than the critical crack size (a_2) that corresponds to the applied stress σ_2. This 'safety' is predicated on the assumptions that the loading is static and that the crack does not grow during service. It is well known, however, that cracks can and do grow in many engineering structures during service by fatigue and/or by stress corrosion cracking. The progressive enlargement of the crack, say from a_1 to a_2, eventually leads to structural failure. While fracture toughness establishes the condition for failure and the (residual) strength of a structural component, its serviceable life or durability is principally a function of the subcritical crack growth resistance of the material (or, for this discussion, its stress corrosion cracking resistance). It is clear, therefore, considerations of stress corrosion crack growth constitutes an important part of materials evaluation and selection, and of structural design.

In this paper, the fracture mechanics appraoch to stress corrosion cracking is briefly summarized. Analytical fracture mechanics considerations, and important considerations in the determination and utilization of stress corrosion cracking data are discussed. Only the case of uniaxial loading, with the loading axis perpendicular to the direction of crack prolongation, will be considered. (Mixed mode loading, although of practical importance, would introduce unnecessary complications into this discussion).

THE FRACTURE MECHANICS APPROACH

The linear fracture mechanics formalism was first applied to the study of stress corrosion crack growth just a little over ten years ago [8,9]. The fracture mechanics approach has undergone considerable development during the ensuing years and has met with considerable success. After overcoming the early objections and resistance, this approach has matured and has become an important element in the consideration of stress corrosion cracking problems, both in terms of research and in terms of design. In essence, the approach recognizes the presence or early iniitiation of cracks in a structural component, and that structural failure results from the growth of these cracks by stress corrosion cracking. The mechanical driving force for this crack growth is considerec to be given by the crack-tip stress intensity factor (K_I) define by linear elasticity. Justification of this approach and for the use of K_I to characterize the mechanical driving force for crack growth has been reviewed by Johnson and Paris [5] and by Wei [6]. Further experimental verification has been provided by the results of Smith et al. [10] and Novak and Rolfe [11].

With the increasing acceptance of the fracture mechanics approach, and the attendant proliferation of new test methodology and terminology, it has become necessary and desirable to establish test standards, as early as practicable, for the orderly development of this important field. ASTM* Committees E-24 on Fracture Testing of Metals and G-1 on Corrosion have jointly undertaken the task for developing fracture mechanics based test methods [12]. In the meantime, it is imperative for the materials engineers and scientists, and designers to be keenly aware of the basic assumptions of the analysis base, and of various other problems associated with stress corrosion testing to ensure proper utilization of test methods and correct interpretation of results.

Fracture mechanics based experimental measurements of stress corrosion cracking susceptibility follow essentially two related approaches, Fig. 2. The choice of a particular approach is determined in part by tradition and motiviation, and in part by practical consideration of experimentation and cost. The simpler, currently more commonly used approach involves the measurement of the time-to-failure (or life), t_F, for precracked specimens under different applied loads (corresponding to different initial K_I), and the determination of a so-called threshold K_I (designated as K_{Iscc}) below which, presumably, no failure can occur as a result of stress corrosion cracking (Fig. 2b) [8,11]. The level of K_{Iscc} in rela-

---

* American Society for Testing and Materials, Philadelphia, PA, USA.

tion to K_{IC}, the plane strain fracture toughness of the material, gives a measure of its stress corrosion cracking susceptibility and is often used in materials selection and design [5]. This approach is analogous to that utilized in conventional stress corrosion testing with smooth or mildly notched specimens, and is widely used in engineering and scientific research at the present time. The other approach is somewhat more complex and involves the determination of the crack growth kinetics, that is, measurements of the rate of crack growth, da/dt, as a function of the mechanical crack driving force, characterized by K_I, under controlled conditions (Fig. 2a). This approach requires greater effort and more sophisticated instrumentation (the basic specimen design being the same for the two approaches). It promises, however, to provide more useful information for quantitative design and life estimation (for fail-safe designs), and for understanding the mechanisms for stress corrosion cracking. The kinetics approach has begun to receive increasing attention in recent years.

Since the approaches and test methods are based on linear fracture mechanics, it is appropriate to review briefly the essential assumptions or linear fracture mechanics analysis. Test methods and specimen design considerations are reviewed in light of these assumptions and of specific problems related to materials response [12].

ANALYTICAL FRACTURE MECHANICS CONSIDERATIONS

Since crack growth and stress corrosion attack would be expected to occur in the highly stressed (strained) region at the crack tip, the stress (or strain) distribution in this region is of primary importance. The stress field near crack tips may be

divided into three basic types, each associated with a local mode
of deformation -- the opening mode (mode I) and two sliding modes
(Modes II and III) [13]. The opening mode (mode I) is characterized
by direct separation of the crack surfaces symmetrically with respect
to the plane occupied by the crack. The edge-sliding mode (Mode II)
and the tearing mode (Mode III) are associated with shear loading
parallel to the crack plane in directions perpendicular and parallel
to the crack front, and are akin to models for edge and screw dislocations respectively. For the present discussion, only mode I
will be considered, although the general discussion will apply to the
other two modes as well.

The near-tip stress and displacement fields associated with
the opening mode (mode I) in an isotropic elastic body are given
by the following equations [1,6,7]:

$$\sigma_x = \frac{K_I}{\sqrt{2\pi r}} \cos \frac{\theta}{2} [1 - \sin \frac{\theta}{2} \sin \frac{3\theta}{2}]$$

$$\sigma_y = \frac{K_I}{\sqrt{2\pi r}} \cos \frac{\theta}{2} [1 + \sin \frac{\theta}{2} \sin \frac{3\theta}{2}]$$

$$\sigma_{xy} = \frac{K_I}{\sqrt{2\pi r}} \sin \frac{\theta}{2} \cos \frac{\theta}{2} \cos \frac{3\theta}{2} \qquad (1)$$

$$u = \frac{K_I}{8\mu} \sqrt{\frac{2r}{\pi}} [(2\kappa - 1) \cos \frac{\theta}{2} - \cos \frac{3\theta}{2}]$$

$$v = \frac{K_I}{8\mu} \sqrt{\frac{2r}{\pi}} [(2\kappa + 1) \sin \frac{\theta}{2} - \sin \frac{3\theta}{2}]$$

For plane strain:

$$\kappa = 3 - 4\nu$$

$$\sigma_z = \nu(\sigma_x + \sigma_y)$$

$$w = 0$$

For generalized plane stress:

$$\kappa = \frac{3 - \nu}{1 + \nu}$$

$$\sigma_z = 0$$

$$w = -\frac{\nu}{E} \int (\sigma_x + \sigma_y)\, dz$$

r and θ are the radial and angular coordinates measured from the crack tip (Fig. 3); μ is the shear modulus; and ν is the Poisson's ratio. Higher order terms in r have been neglected in Eq. 1. Hence, the stresses and displacements are to be regarded as good approximations in the region where r is small compared to the other planar (x-y plane) dimensions, such as crack length, and exact in the limit as r approaches zero.

The parameter K_I is the stress intensity factor for mode I, which depends on the loading and the configuration of the body, including the crack size, and governs the intensity or magnitude of the local stresses. The analytical determination of K_I and the crack-tip stress fields is basically a problem in the mathematical theory of elasticity and has been summarized elsewhere (see, for example, [15,16]). Stress intensity factors for many different loading conditions and body configurations have been catalogued by Sih [15] and by Tada et al. [17]. Numerical solutions of K_I for practical test specimens, suitable for routine use, are also given in these references [15,17] and in ASTM STP 410 [3].

It should be noted that the linear elasticity solution for a sharp crack (see Eq. 1 for example) gives rise to infinite stresses at the crack tip where the radius of curvature is 'zero'. In reality, of course, the deformed shape of the crack would assume some finite radius of curvature, and the stresses would always remain finite. Hence, it is likely that a large deformation theory would predict finite stresses at the crack tip [18]. In addition, the occurrence of local plastic deformation also tends to reduce the stress concentrating effects of the crack. If the zone of plastic deformation is small in comparison with the crack length and other planar dimensions of the body, then the stress distribution in the large will not be seriously disturbed and the elasticity solutions represent a reasonably accurate approximation of the stress and displacement fields near the crack tip. Since the small zone of plastically deformed material at the crack tip is contained within the surrounding elastic material, it is reasonable to expect that the behavior in this region would be governed by the surrounding elastic material and thus be characterized by the crack-tip stress intensity factor K_I.

Inspection of Eq. 1 suggests that identical stress fields are obtained for identical K_I values. Hence, K_I provides a single parameter characterization of the stress and displacement fields near the crack tip, independent of the size of the crack. This is not strictly true, however, since Eq. 1 is only approximate for large values of r. The exact analysis of Inglis [19] shows that for cracks of different sizes, for the same K_I, the stresses away from the crack tip differ considerably; the stresses very close to the crack tip, though, are nearly equal to each other [20]. On the basis of the Inglis solution [19], Liu [20] showed that if the plastically deformed zone is much smaller than some region, within which the elastic stresses are approximately the same for bodies with different sizes of crack loaded to the same level of K_I, the stresses and strains at geometrically similar points, even within the plastic zone, would be the same. Thus, the use of K_I as a single parameter characterization of the crack driving force may be justified. For stress corrosion cracking studies, the stress intensity factor for the opening mode (mode I) of crack growth (K_I) is generally used, since the opening mode predominates in stress corrosion cracking.

It must be emphasized that the use of the crack-tip stress intensity K_I, defined by linear elasticity, to characterize the mechanical crack driving force in practical engineering materials is predicated on the assumption of limited plasticity. The applicability of this approach to stress corrosion cracking and other fracture studies depends on the experimental fulfillment of this fundamental assumption. The foregoing discussions are also limited to the case of stationary cracks, in contradistinction to moving cracks that are of concern in subcritical-crack growth (stress corrosion crack growth, in particular). For the usual rates of crack growth encountered in subcritical-crack growth studies, however, modification of K_I to include dynamic effects will not be necessary [21,22]

TEST METHODS AND SPECIMEN DESIGN

Various types of specimens and methods of loading can be used to determine the stress corrosion cracking properties of materials -- that is, either the crack growth kinetics or K_{Iscc}. The fracture mechanics based specimens may be broadly separated into three groups:

- Constant load, increasing K_I specimens
- Constant displacement, decreasing K_I specimens
- Constant K_I specimens

The first group of specimens is exemplified by the cantilever bend specimens [8]; the second, by the bolt loaded WOL (wedge-opening-load) specimens [11,23]; and the third, by tapered DCB (double-cantilever-bend) specimens [24] subjected to constant applied load. Of course, by judicious placement of the applied loads and choice of loading conditions, increasing or decreasing K_I conditions may be obtained with any of the specimen groups. Specimens of the first and second groups have been commonly used for kinetics studies and K_{Iscc} determinations, and are being considered for use in the test methods under development by ASTM Committees E-24 and G-1. Specimen size considerations and problems associated with the use of these types of specimens have been discussed in detail previously [6,12], and will be summarized in this and later sections.

In terms of specimen size, there are two distinctly separate requirements -- one pertaining to the applicability of the linear elastic analysis, and the other with regard to the condtion of constraint at the crack tip (that is, plane strain versus plane stress). The first of these requirements relates to the minimum size of crack and of other planar dimensions of the specimen that are needed to satisfy the assumptions of limited plasticity. The second one refers to the degree of relief of constraint in the thickness direction by localized plastic deformation (yielding) at the crack tip. Both of these requirements relate to the size of the crack-tip plastic zone, which in itself is not adequately defined. These requirements, therefore, cannot be predicted from theoretical considerations alone [3,6], and must be established by trial. Experimental data from fracture toughness tests have provided some useful guidelines [3]. It has been convenient to use the parameter $(K_I/\sigma_{ys})^2$ as a measure for the size of the plastic zone at the crack tip, where σ_{ys} is taken to be the uniaxial tensile yield strength [3]. Fracture toughness test data suggest that to satisfy the assumption of limited plasticity, the minimum crack length should be equal to or greater than 2.5 $(K_I/\sigma_{ys})^2$. Furthermore, to provide reasonable ranges of K_I and of crack growth, and for best accuracy of K_I in practical specimens, the range of crack lengths should be between one-quarter to three-quarters of the specimen width (that is $0.25W \leqslant a \leqslant 0.75W$). Taking the initial crack length (a_o) to correspond with the lower of two values, the following dimensional requirements are obtained and are not unreasonable [3,6]:

$$a_o \geq 2.5 \ (K_I/\sigma_{ys})^2 \tag{2}$$

$$W = 4a_o \geq 10 \ (K_I/\sigma_{ys})^2 \tag{3}$$

Typically, the maximum K_I to be encountered during a test is used in defining the specimen size, although some adjustments are usually made in practice to accommodate to specific ranges of K_I values and to other test requirements. The specimen thickness that is required to achieve conditions approximating plane strain is less clear. In plane strain fracture toughness testing, a thickness equal to or greater than 2.5 $(K_{Ic}/\sigma_{ys})^2$ is used. This condition is only applicable at the onset of crack growth instability from a 'pre-crack' (or, more commonly, at 'pop-in'), and has been mistakenly used often as the plane strain criterion in the literature. Certainly, the specimen thickness should be much greater than 2.5 $(K_I/\sigma_{ys})^2$. The precise requirement, however, remains to be established by adequate experimentation. For practical applications, it is the usual practice to test materials in the thicknesses that will be employed in service.

In preparing a test specimen for stress corrosion testing, one of two precracking procedures have been utilized. In the first procedure, a crack is extended from the starter notch by fatigue (cyclic loading). The maximum K_I level is usually kept to well

below the starting K_I for the test or the anticipated K_{Iscc}. In
the second procedure, the starting crack is initiated from the
starter notch by monotonically increasing the applied load (the
so-called 'pop-in' procedure). The first procedure is preferred
in that the deformation is much more localized and the starter
crack more closely resembles those encountered in service [3].
The second procedure is more commonly used with the 'decreasing
K_I' tests. Large deformations accompanies crack iniitiation in
this case. Their effects on the crack growth kinetics and on
K_{Iscc} have not been clearly established, although some influence
can certainly be expected [25,26]. As such, the use of this procedure for precracking is not advisable at this time.

Several quantitative methods are available for monitoring
crack growth [27-32]. Crack length measurements are most commonly made by visual observations. They may be inferred also from
changes in specimen compliance [27] or electrical resistance (or
potential) [28-30]. Alternatively, crack lengths may be determined by ultrasonic or eddy-current techniques [31,32]. Each of
these techniques has its inherent advantages and disadvantages.
A thorough evaluation of the technique should be made prior to its
adoption for use.

KINETICS OF CRACK GROWTH AND LIFE (TIME-TO-FAILURE) APPROACHES

In studying crack growth under stress corrosion conditions,
it has been generally assumed that, under constant environmental
conditions, the rate of crack growth depends only on the mechanical
crack driving force K_I, and is invariant with respect to time for
constant K_I. This is certainly the case for steady-state crack
growth, and must be so if K_I is to be a proper representation of

the mechanical crack driving force. This correspondence, however, does not preclude the occurrence of a number of nonsteady-state (time-dependent) phenomena that can also depend on K_I, as discussed by Wei et al. [12]. Close examination of the crack growth process showed that crack growth, in fact, occurs in six stages [12].

- Crack growth on rising load
- Initial transient crack growth
- Incubation period*
- Crack acceleration
- Steady-state crack growth
- Onset of failure or crack growth instability

The occurrence of crack extension on loading, followed by transient growth, has been observed by several investigators and is illustrated in Fig. 4 [12,33-35]. Following the initial transient growth, the crack appears to stop growing, or enter an incubation period, before accelerating to some steady-state rate of growth appropriate to the prevailing K_I. The entire nonsteady-state growth period (involving the first four stages) is a function of K_I and test temperature [12,35]. The incubation period and the period of crack acceleration are illustrated more clearly in Fig. 5** [12]. The existence of an incubation period for precracked specimens has been demonstrated by Benjamin and Steigerwald [36] on AISI 4340 steel tested in water. They showed that the incubation period is affected by prior history. The incubation period depends also

---

* Incubation period is arbitrarily defined as that period during which the rate of crack growth is less than 10^{-6} in./min (4×10^{-10} m./sec).

** Crack growth was monitored by a displacement gage. The oscillatory nature of the steady-state growth rate shown in Fig. 5 is principally an artifact introduced by changes in displacement produced by oscillations in the dead weight loading device.

on K_I and on the alloy-environment system, and can vary from less than one minute to several thousand hours [12].

Typical steady-state crack growth response as a function of K_I is illustrated by the results of Williams [37], Fig. 6, and is also shown schematically in Fig. 2a. Steady-state crack growth kinetics may be divided into three stages, Fig. 2a. Stage I is highly dependent on K_I, and may reflect crack acceleration for certain types of tests. Stage II in independent or nearly independent of K_I and represents a range where crack growth is rate limited by the embrittling chemical process. In Stage III, the condition of unstable crack growth is approached, and the rate becomes again highly dependent on K_I. For high strength materials, under conditions approximating plane strain, the condtion for the onset of unstable growth is defined by $K_I = K_{Ic}$, where K_{Ic} is the plane strain fracture toughness of the material. In high strength materials of relatively low toughness, Stage II crack growth may be suppressed by the early onset of instability (see Fig. 7 for example [38]). In these cases, a pronounced Stage II would not be observed, and care must be taken in drawing mechanistic inferences from the experimental data.

The life of a specimen is a function of both the incubation and crack growth processes. A typical time-to-failure (life) versus K_{Ii} curve for a 'constant load, increasing K_I' specimen is illustrated schematically in Fig. 2b, where K_{Ii} denotes the initial applied stress intensity factor. The time-to-failure (t_F) is composed of an incubation period (t_{INC}) and a period of slow crack growth (t_{SC}) [12].

$$t_F = t_{INC} + t_{SC} \tag{4}$$

The incubation time is a function of K_I and prior history. The

period of slow crack growth depends on the specimen configuration
and the details of the crack growth kinetics [6,12,37]. The time-
to-failure is related inversely to the rate of crack growth,
Fig. 2, and may be estimated from the kinetic data. The general
form of the crack growth kinetics may be expressed as follows:

$$\frac{da}{dt} = G \begin{array}{l} (K_I, K_{Ii}, T, \text{ and other material} \\ \text{environment, and test variables}). \end{array} \qquad (5)$$

where T is the test temperature. The inclusion of K_{Ii}, the initial
stress intensity factor, reflects the recognition that a steady-
state rate of crack growth may not be established immediately on
loading. Thus, da/dt can be dependent on time for a given K_{Ii}.
For steady-state crack growth , da/dt becomes simply a function of
K_I and of variables other than K_{Ii}, and is therefore independent
of time.

$$\frac{da}{dt} = F(K_I, T, \ldots) \ldots \qquad (6)$$

Under the assumption of steady-state crack growth, the time-to-failure
for a typical test specimen may be obtained by direct integration,
when the applied load and all of the other variables are maintained
constant. For constant load tests:

$$\frac{dK_I}{dt} = \frac{dK_I}{da}\frac{da}{dt} = \frac{dK_I}{da} F \ldots \qquad (7)$$

Since, for the steady-state case, the rate of crack growth is time
independedt, Eq. 7 may be integrated directly. Separation of variables
and integration gives the time-to-failure as:

$$t_F = t_{INC} + \int_{t_{INC}}^{t_F} dt = t_{INC}$$

$$+ \int_{K_{Ii}}^{K_{Ic}} \left[\frac{dK_I}{da} \cdot F\right]^{-1} dK_I \qquad (8)$$

The incubation time t_{INC} is a function of K_{Ii}, the initial stress intensity factor. K_{Ic} is the plane strain fracture toughness of the material. The stress intensity factor K_I can be expressed in the form [3]:

$$K_I = \sigma a^{1/2} \, Y \left(\frac{a}{W}\right)\ldots \qquad (9)$$

where σ = the nominal applied stress, a = crack length, $Y(a/W)$ = a paramenter representing the crack and specimen geometries, and W = the specimen width. Inspection of Eqs. 8 and 9 indicates that the time-to-failure will depend on the specimen geometry and size. For example, the time-to-failure for geometrically similar specimens loaded to identical K_{Ii} levels is expected to be different. Similarly, specimens with different initial crack sizes, loaded to the same initial K_I, will produce different lives. Typical time-to-failure curves computed on the basis of crack growth kinetics and assumed crack geometries are shown in Figs. 8 and 9 [37]. Even though the incubation time and nonsteady-state crack growth were neglected in these calcualtions, the essential features of the t_F versus K_{Ii} curves are reproduced (see Fig. 2b). It is clearly seen that the time-to-failure is related to the crack growth kinetics and that it is dependent on geometry and environmental conditions [12].

IMPLICATIONS FOR STRESS CORROSION CRACKING STUDIES

In the foregoing discussion, it has been shown that the steady-state rate of stress corrosion crack growth is uniquely related to the crack driving force, with other conditions being constant. The mechanical component of the crack driving force may be characterized, under the assumption of limited plasticity, by the crack tip stress intensity factor, K_I, defined by linear elastic fracture mechanics. The kinetic information can be quite useful in making quantitative estimates of the service lives of structural components provided that the incubation time and the period of nonsteady-state crack growth can be handled in some satisfactory way. The existence of an incubation period and nonsteady-state crack growth presents serious practical problems in the determination and utilization of crack growth kinetics. Figures 5 and 10 show that the incubation phenomenon and nonsteady-state crack growth can lead to an under-estimation of the steady-state rate of crack growth, with a con-sequent overestimation of the safe operating life.

The incubation phenomenon and the crack growth kinetics (for both the steady- and nonsteady-state cases) can affect the evalua-tion and use of the so-called threshold stress intensity factor for stress corrosion cracking, K_{Iscc}, as well. Kinetic data are some-times used for estimating K_{Iscc}. If the apparent rapid decrease in the rate of crack growth shown in Fig. 10 is interpreted as an approach to the threshold K_I level, erroneously high estimates of K_{Iscc} would result. The estimated value of K_{Iscc} would appear to depend on the starting K_I level (K_{Ii}) used in testing. For the usual stress corrosion cracking tests (that is, constant-load, increasing K_I, tests for K_{Iscc}), some type of criteria are nor-mally used to terminate the test and to define an estimated value of K_{Iscc}. In practice, K_{Iscc} is defined as that K_I level at which 'no failure' or 'no observable crack growth' has occurred after some prescribed period of time. Since it has been shown through con-sideration of the crack growth kinetics that the time-to-failure is quite dependent on the loading condition, specimen size and geometry, and environmental conditions, such criteria can lead to

serious errors in the estimation of K_{Iscc}. The problem is com-
pounded by the existence of incubation. For example, by increasing
the cut-off time from 100 to 10,000 hours (or, from approximately
4 days to over 1 year), the apparent K_{Iscc} is decreased from
170 ksi√in to 25 ksi√in (187 to 27.5 MN-m$^{-3/2}$) for a high alloy
steel tested in synthetic sea water at room temperature [12]. Thus,
substantial error can be introduced by using short-time test data in
design. Because the apparent K_{Iscc} is so dependent on test procedures and conditions, its practical utility must be carefully
re-evaluated.

In developing test specifications, one must be certain that the
incubation period and the nonsteady-state crack growth processes
are reduced or eliminated in experimentation, or taken into proper
consideration in data reduction and interpretation. The influence
of the kinetics of steady-state crack growth should also be considered. For kinetic studies, constant K specimens may be used.
The testing time at each K_I level, however, must be sufficiently
long to ensure the establishment of steady-state conditions. The
incubation time at low K_I levels may be too long to justify the use
of this test method. Since the incubation time is expected to be
much shorter at high K_I levels, the use of a 'constant displacement,
decreasing K' specimen may be more attractive [12]. Experience
with this type of specimen for kinetic studies is limited at this
time; further development will be required. For K_{Iscc} determinations,
this type of specimen has been used and offers definite advantages.
By starting at high K_I levels, it is expected that the long incubation times can be avoided. Consideration of the proper cut-off
time, however, must still be established. Design of the specimen
must be such that the decrease in K_I with crack prolongation is not

84

too rapid. A rapid decrease may produce delays in crack growth or
exhibit the prestressing effect [25,26], with consequent overestimation in K_{Iscc} [12].

The existence of a K-independent stage (Stage II) of crack
growth suggests that crack growth in this stage is rate limited
by the operative chemical process for stress corrosion cracking.
This realization has provided the essential link between the
mechanical and chemical aspects of stress corrosion cracking, and
represents a most important contribution of linear fracture mechanics. By examining the responses of Stage II growth rates as

functions of temperature and pressure (or concentration of corrosive species), and comparing them to those for candidate chemical
processes, it is now possible (at least in principle) to deduce the
rate controlling processes and mechanisms for stress corrosion
cracking. Such mechanistic studies require close interactions between
researchers in fracture mechanics, chemistry, and materials science,
and are to be encouraged.

REFERENCES

1. Irwin, G. R., in Sturctural Mechanics, Pergamon Press, 557
    (1960).

2. Fracture Toughness Testing and Its Applications, ASTM STP
    381, American Society for Testing and Materials (1965).

3. Brown, W. F., Jr. and Srawley, J.E., Plane Strain Crack
    Toughness Testing of High Strength Metallic Materials, ASTM
    STP 410, American Society for Testing and Materials (1966).

4. Sih, G. C., in Methods of Analysis and Solutions of Crack
    Problems, G. C. Sih, Ed., Noordhoff, XXI - XLV (1973).

5. Johnson, H. H. and Paris, P. C., Journal of Engineering
    Fracture Mechanics 1, 3-45 (1968).

6. Wei, R. P., in Proceedings of Conference - Fundamental Aspects of Stress Corrosion Cracking, National Association of Corrosion Engineers 104 (1969).

7. Wei, R. P., Journal of Engineering Fracture Mechanics 1, 633 (1970).

8. Brown, B. F. and Beachem, C. D., Corrosion Science 5, 745-750 (1965).

9. Johnson, H.H. and Willner, A. M., Applied Materials Research, 34 (1965).

10. Smith, H.R., Piper, D.E. and Downey, F.K., Journal of Engineering Fracture Mechanics 1, 123-128 (1968).

11. Novak, S.R. and Rolfe, S.T., Corrosion 26, 121-130 (1970).

12. Wei, R.P., Novak, S.R. and Williams, D.P., Materials Research and Standards, ASTM, 12, 25-30 (September 1972); also in AGARD-CP-98 (1972).

13. Paris, P.C. and Sih, G.C., in Fracture Toughness Testing and Its Applications, ASTM STP 381, American Society for Testing and Materials, 30-38 (1965).

14. Irwin, G.R., Journal of Applied Mechanics, Transactions of the American Society of Mechanical Engineers 24, 36, (1957).

15. Sih, G.C., Ed., Methods of Analysis and Solutions of Crack Problems, Noordhoff (1973).

16. Sih, G.C., Handbook of Stress Intensity Factors, Institute of Fracture and Solid Mechanics, Lehigh University, Bethlehem, Pennsylvania (1973).

17. Tada, H., Paris, P.C. and Irwin, G.R., Stress Analysis of Crack Handbook, Del Research Corporation, Hellertown, Pennsylvania (1973).

18. Sih, G.C. and Liebowitz, H., in Fracture - An Advanced Treatise, Vol. II, H. Liebowitz, Ed., Academic Press (1970).

19. Inglis, E.E., Transactions, Institution of Naval Architects, (London) 60, 219 (1913).

20. Liu, H.W., Discussion, Fracture Toughness Testing and Its Applications, ASTM STP 381, American Society for Testing and Materials, 23 (1965).

21. Sih, G.C., International Journal of Fracture Mechanics 4 (1968)

22. Chou, Y.T. and Wei, R.P., Scripta Metallurgica 6, 965 (1972).

23. Novak, S.R. and Rolfe, S.T., Journal of Materials, ASTM 4, 701 (1969).

24. Marcus, H.L. and Sih, G.C., Journal of Engineering Fracture Mechanics 3, 453 (1971).

25. Carter, C.S., Metallurgical Transactions 3, 584 (1972).

26. Jonás, O., Corrosion 29, 299-304 (1973).

27. Brown, B.F., Ed., Stress Corrosion Cracking in High Strength Steels and in Titanium and Aluminum Alloys, Naval Research Laboratory, Washington, D.C. (1972).

28. Johnson, H.H., Materials Research and Standards, ASTM 4, 422 (1965).

29. Li, Che-Yu and Wei, R.P., Materials Research and Standards, ASTM 6, 392 (1966).

30. McIntyre, P. and Priest, A.H., British Steel Corporation, Open Report No. MG/54/71 (1971).

31. Clark, W.G., Jr., Journal of Engineering Fracture Mechanics, 1, 385 (1968).

32. Swanson, S.R., Cicci, F. and Hoppe, W., in Fatigue Crack Propagation, ASTM STP 415, American Society for Testing and Materials, 312-362 (1967).

33. Li, Che-Yu, Talda, P.M. and Wei, R.P., unpublished results (1965).

34. Barsom, J.M., unpublished results (1966).

35. Landes, J.D. and Wei, R.P., International Journal of Fracture 9, 277 (1973).

36. Benjamin, W.D. and Steigerwald, E.A., Transactions of the American Society for Metals 60, 547-548 (1967).

37. Williams, D.P., International Journal of Fracture 9, 63-74 (1973).

38. Miller, G.A., Hudak, S.J., Jr. and Wei, R.P., Journal of Testing and Evaluation, ASTM 1, 524 (1973).

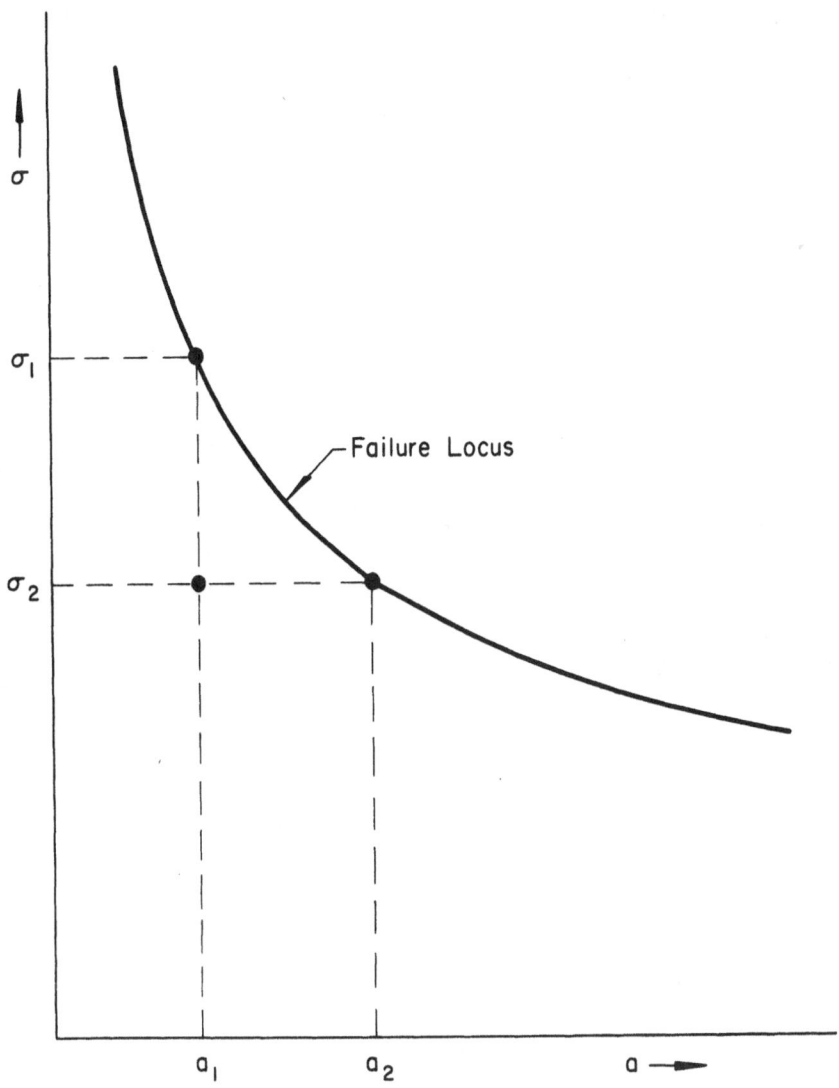

Fig. 1:   Schematic Representation of Failure Locus.

88

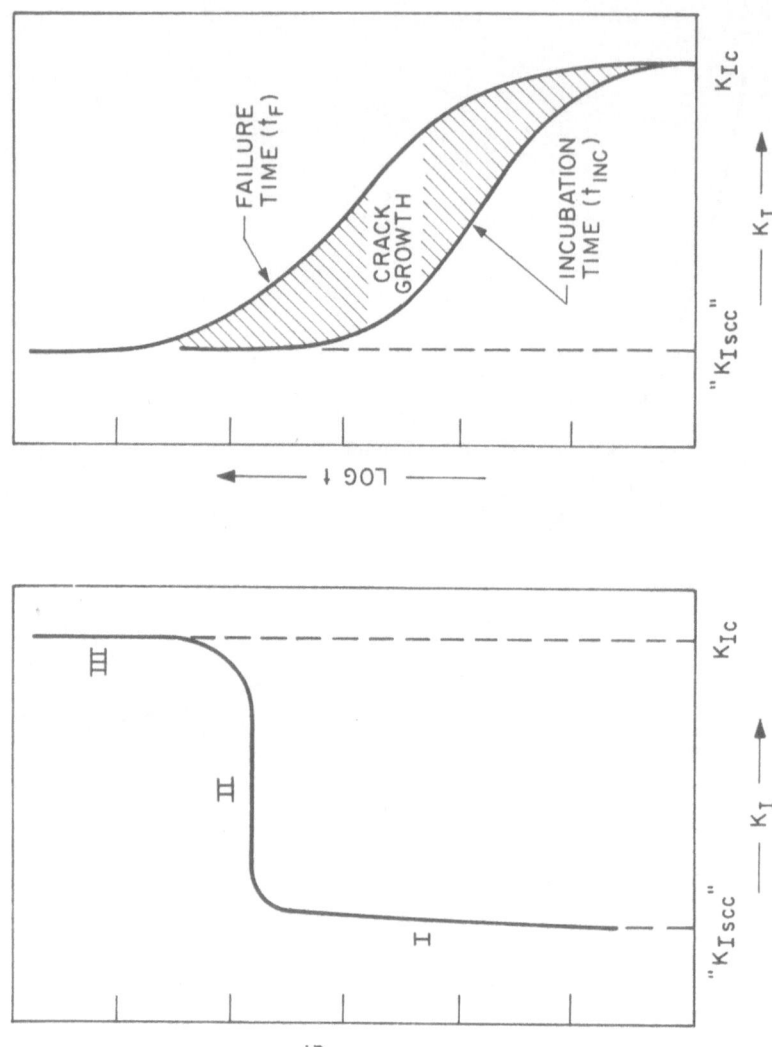

Fig. 2: Schematic Representations of the Crack Growth Kinetics and Time-To-Failure under Sustained Loads [12].

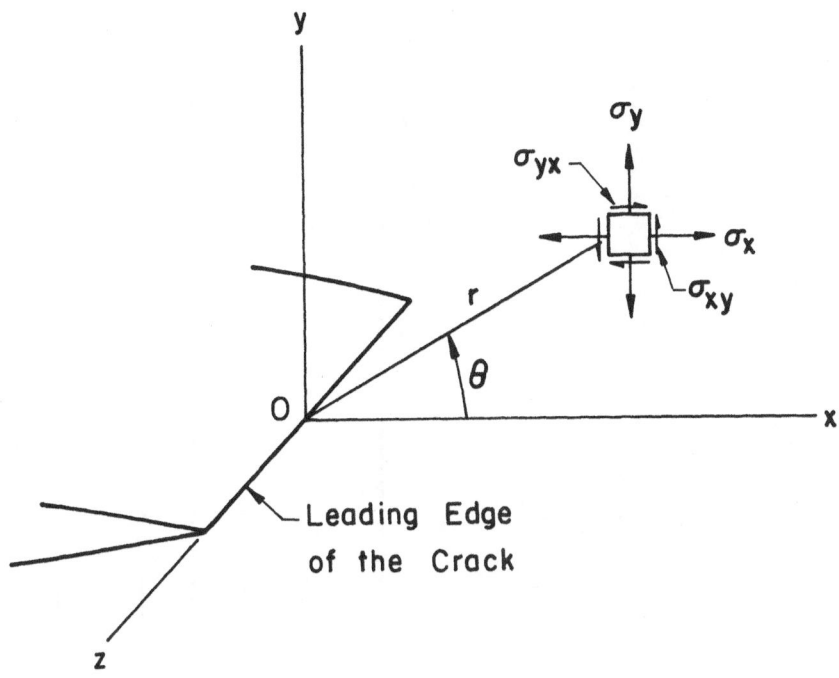

Fig. 3: Coordinates and Stress Components in the Crack-Tip Stress Field.

90

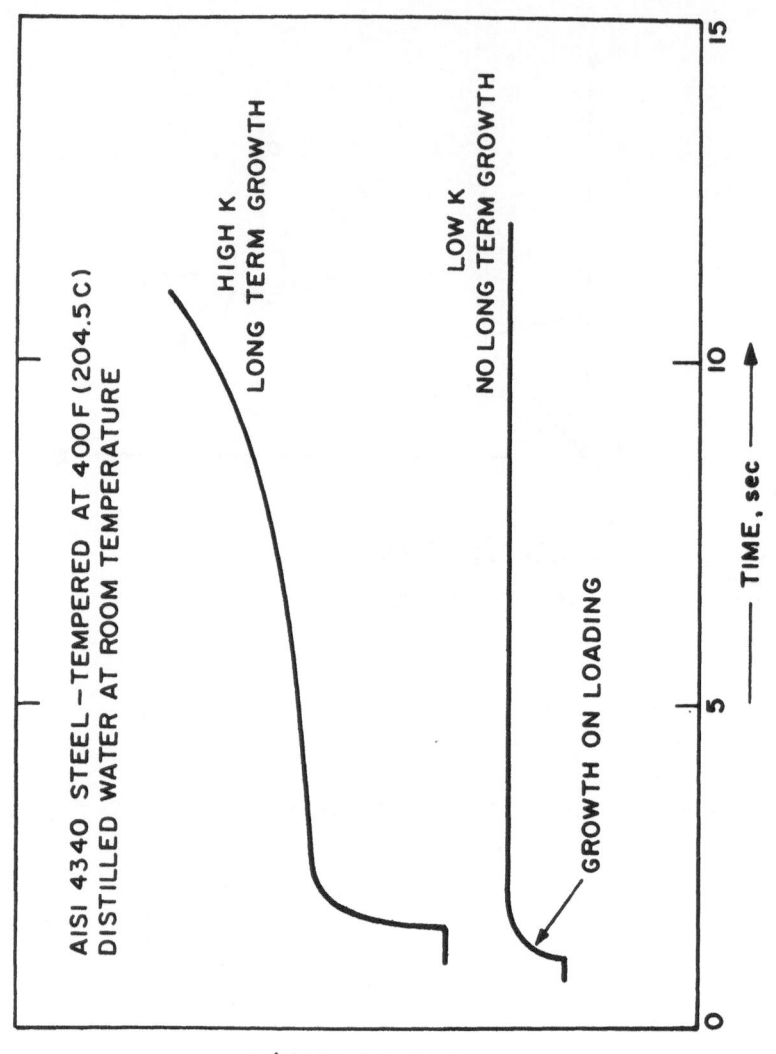

Fig. 4: Schematic Illustration of Sustained Load Crack Growth Behavior [35].

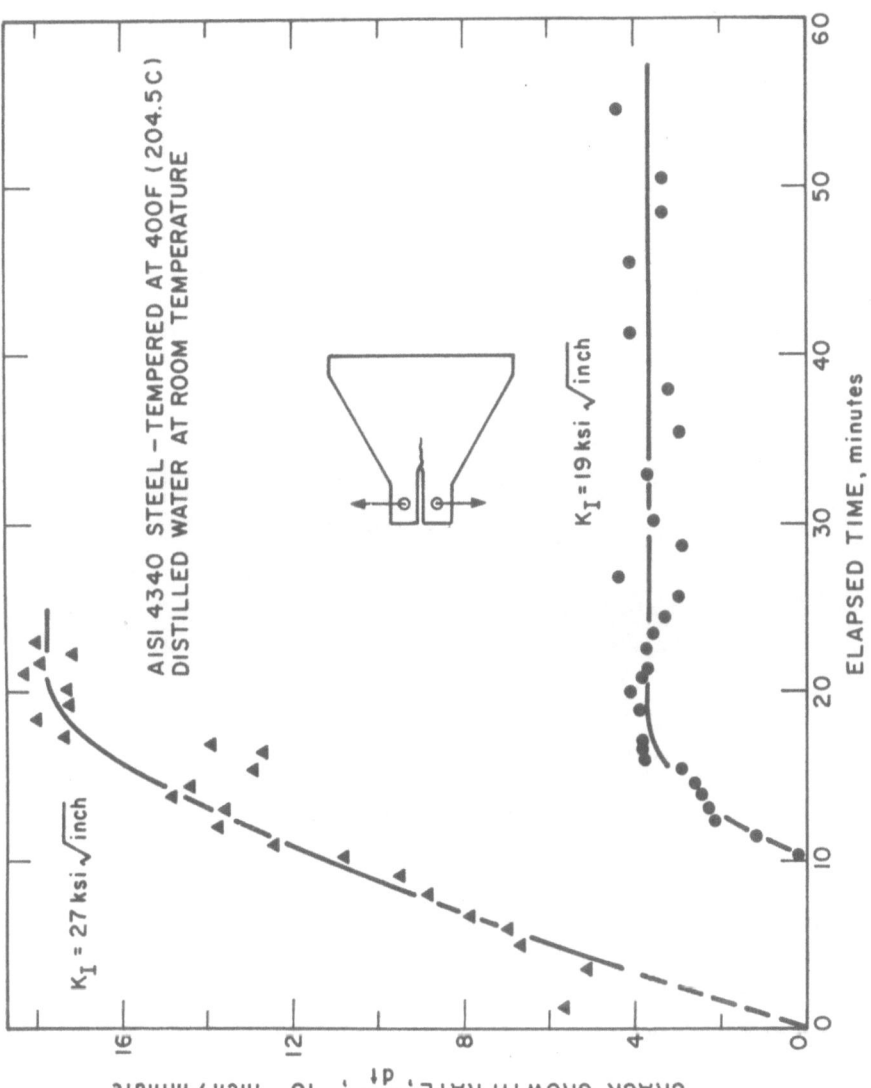

Fig. 5: Sustained Load Crack Growth under Constant K_I Showing
Incubation, Crack Acceleration, and Steady-State
Stages of Crack Growth [12].

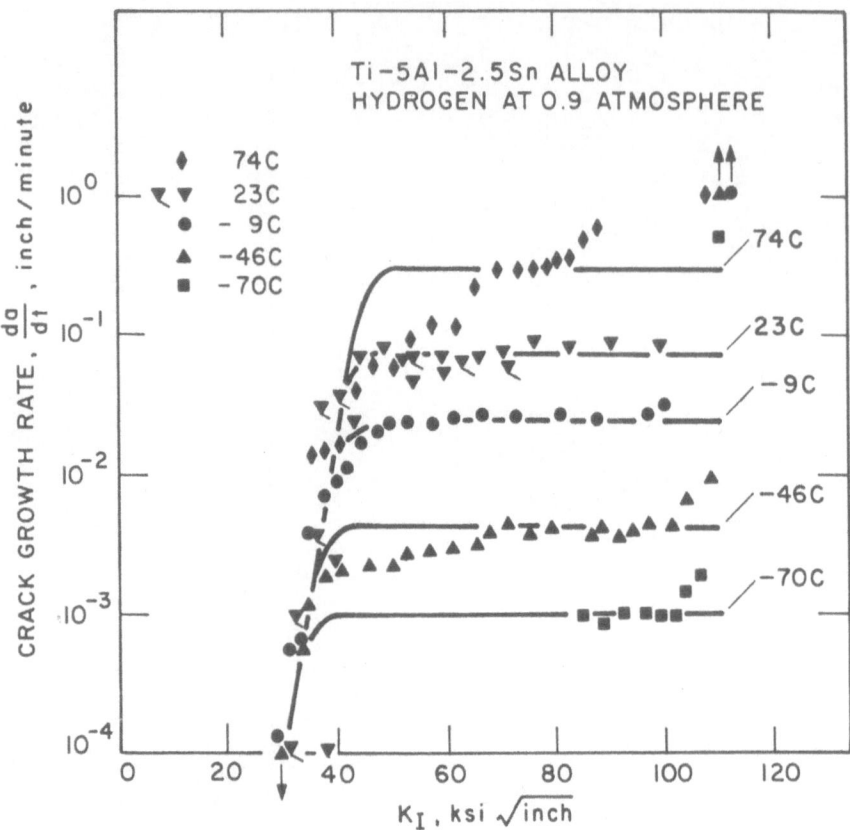

Fig. 6: Typical Steady-State Crack Growth Kinetics [37].

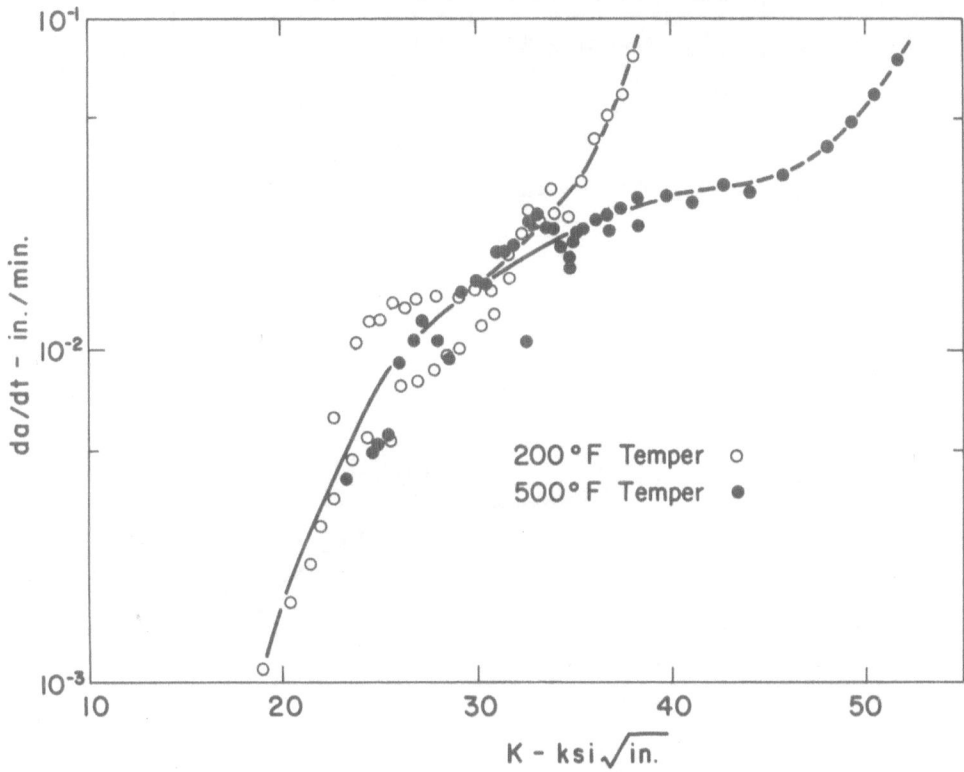

Fig. 7: The Kinetics of Sustained-Load Crack Growth in Distilled
Water at Room Temperature for AISI 4340 Steels [38].

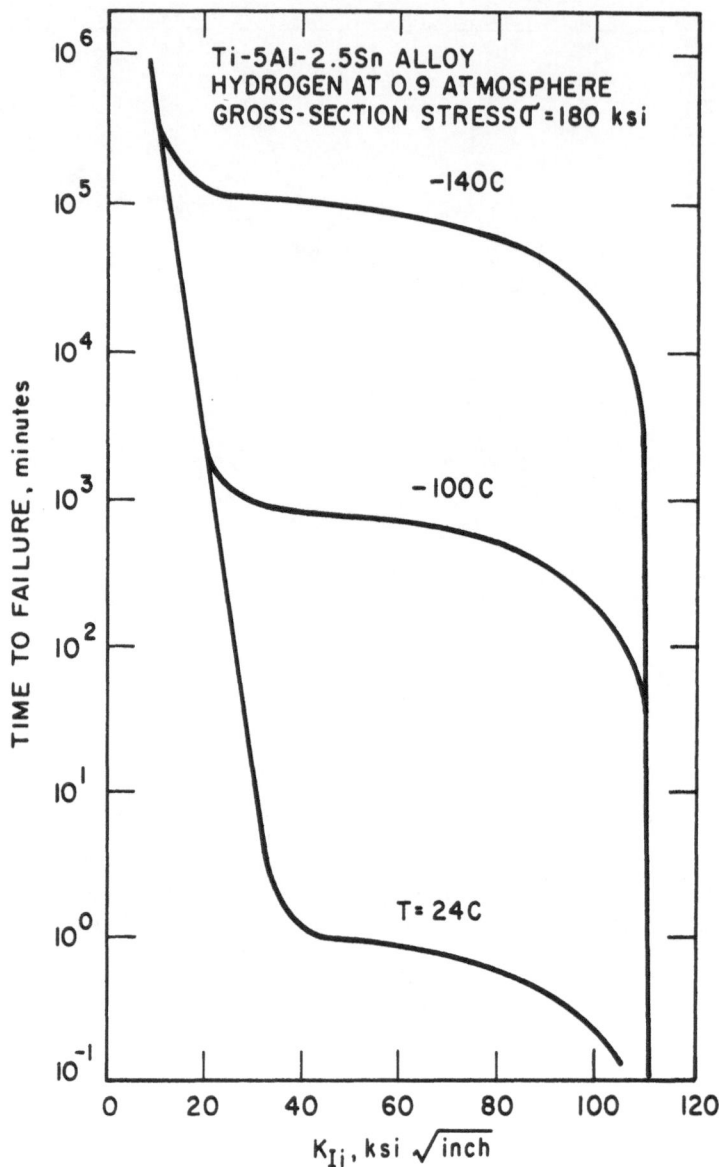

Fig. 8: Computed Time-To-Failure Curves (Excluding Incubation Time) Showing the Influence of Test Temperature [37].

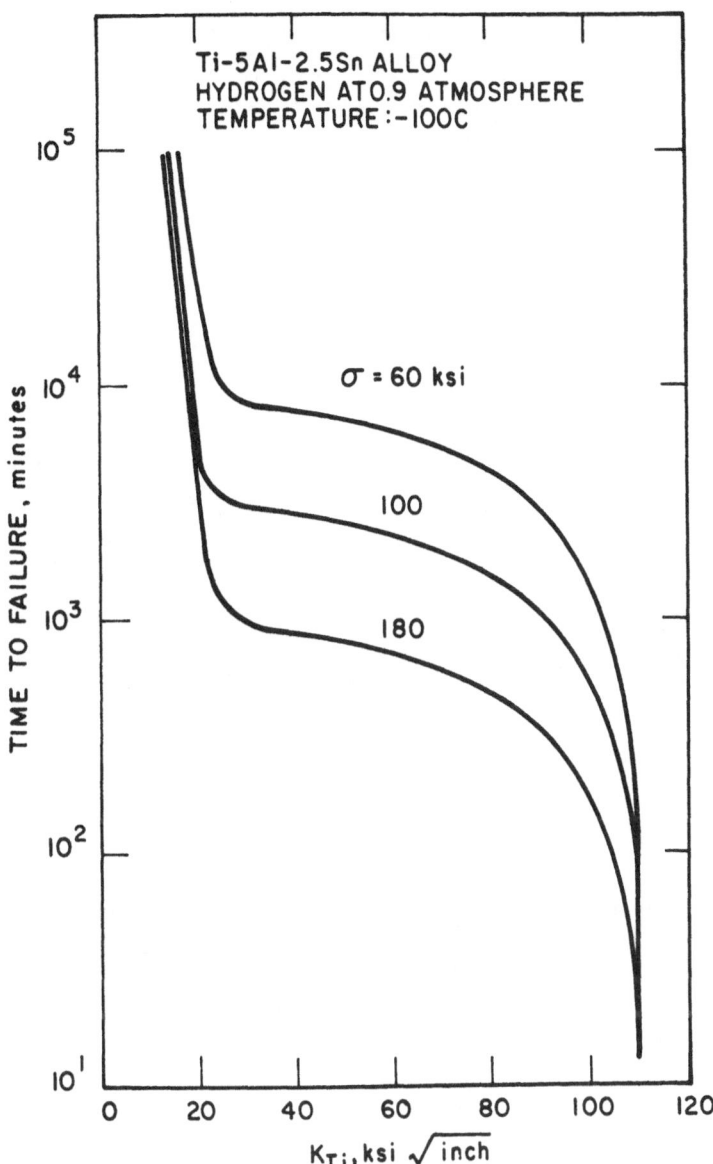

Fig. 9: Computed Time-To-Failure Curves (Excluding Incubation
Time)Showing the Effect of Gross Section Stress or
Initial Crack Size [37].

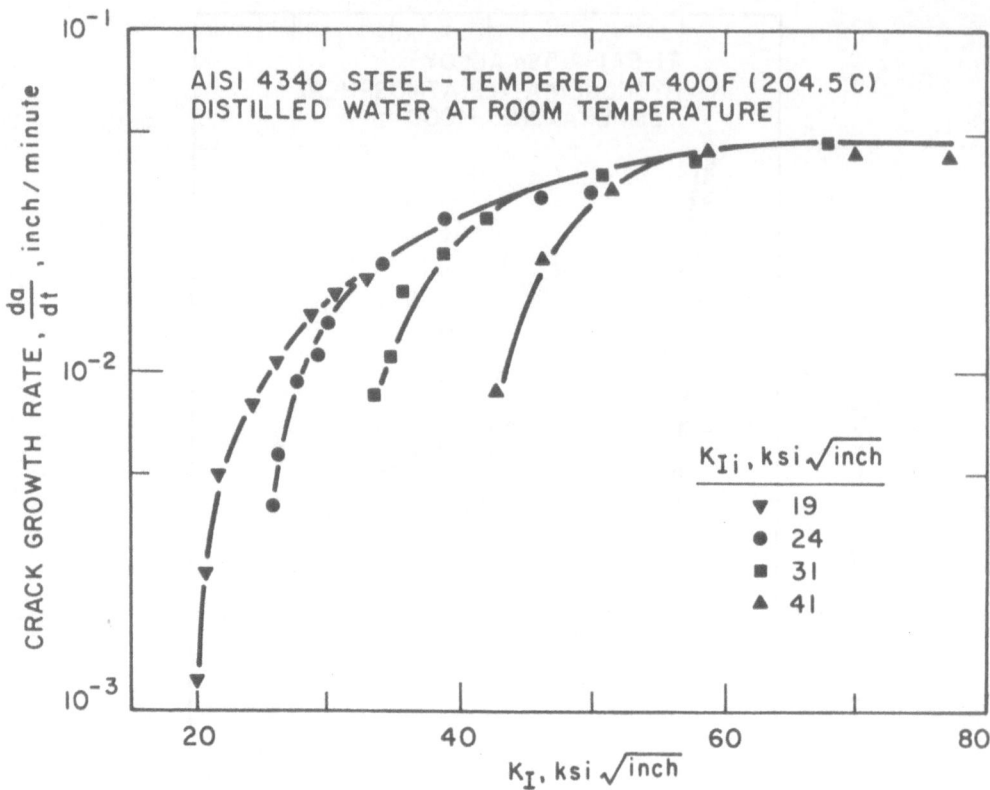

Fig. 10: Kinetics of Sustained Load Crack Growth Showing the
          Effect of Initial K_I [35].

STRESS CORROSION CRACKING OF CAST ALUMINUM ALLOYS

Markus O. Speidel[*]

Department of Metallurgical Engineering
Ohio State University, Columbus, Ohio USA

Cast aluminum alloys are not as widely used as wrought
aluminum alloys, especially in high-strength applications
where stress corrosion cracking may be a problem with the
wrought alloys[1]. Thus, neither the stress corrosion performance, nor the composition, nor the designation, nor
the mechanical properties, nor the heat treatment of cast
aluminum alloys are as widely known as those of wrought
aluminum alloys. It appears, however, worthwhile to outline
these topics briefly because improved alloys and heat
treatments are now available which result in a combination
of high strength and high resistance to stress corrosion
cracking in cast aluminum alloys.

Cast Aluminum Alloy Designation and Composition

The unified numbering system (UNS) for cast aluminum alloys
consists of the prefix A (for aluminum) followed by five
digits[2]. The last four of these digits are identical with
the numbering system the Aluminum Association has established

---

[*] On leave of absence from Brown Boveri Research Center,
CH-5401 Baden, Switzerland

to identify cast aluminum alloys (AA)[3]. Of these four digits the first one indicates the alloy group as follows:

Aluminum 99.0% minimum 1XX.X
Copper.......................................2XX.X
Silicon, with added Copper and/or Magnesium..3XX.X
Silicon......................................4XX.X
Magnesium....................................5XX.X
Zinc...6XX.X
Tin..8XX.X
Other Element................................9XX.X

The second two digits identify the aluminum alloy. The last digit, which is separated from the others by a decimal point, indicates the product form; castings carry the number 0. A modification of the original alloy is indicated by the letter before the numerical disignation. Experimental alloys are indicated by the prefix X. Table 1 presents alloy designations and chemical composition limits for a number of commonly used aluminum casting alloys. There are many more aluminum casting alloys available than those listed in Table 1. Further information is contained in the references [1] through [8]

Table 1 provides also an overview over the composition limits of many widely used cast aluminum alloys. Further information concerning other cast aluminum alloys can be obtained from the Aluminum Association or from foundries. The alloys listed in Table 1 include those for which stress corrosion information is available[1,5,9-20].

Heat Treatment and Mechanical Properties of Cast Aluminum Alloys

A temper designation for cast aluminum alloys has been devised by the Aluminum Assocation. This designation system is identical to the one for wrought aluminum alloys. For example, alloys in the as cast condition are marked -F. -T6 is aged to near peak strength, and -T7 is overaged to achieve certain other properties, such as improved resistance to stress corrosion cracking. Examples of heat treatments for cast aluminum alloys are given in Table 2.

The mechanical properties of cast aluminum alloys depend not only on the heat treatment after casting, but also on the product form (e.g. sand cast versus permanent mold cast) and on a number of processing variables such as section thickness, cooling rate, and so on.

Table I Designation and Composition of some Cast Aluminum Alloys

Alloy Designation UNS	AA	Al	Cu	Fe	Mg	Mn	Si	Ti	Zn	(other)	other,each	other, total
A02010	201.0	rem	4.0-5.2	0.15 max	0.15-0.55	0.20-0.50	0.10 max	0.15-0.35	--	Ag 0.40 - 1.2	0.05 max	0.10 max
A02240	224.0	rem	4.5-5.5	0.10 max	--	0.20-0.50	0.06 max	0.35 max	--	V 0.05-0.15, Zr 0.10-0.15	0.03 max	0.10 max
A02490	249.0	rem	3.8-4.6	0.10 max	0.25-0.50	0.25-0.50	0.05 max	0.02-0.35	2.5-3.5	--	0.03 max	0.10 max
A02950	295.0	rem	4.0-5.0	1.0 max	0.03 max	0.35 max	0.7-1.5	0.25 max	0.35 max	--	0.05 max	0.15 max
A03190	319.0	rem	3.0-4.0	1.0 max	0.10 max	0.50 max	5.5-6.5	0.25 max	1.0 max	--	--	0.50 max
A03330	333.0	rem	3.0-4.0	1.0 max	0.05-0.50	0.50 max	8.0-10.0	0.25 max	1.0 max	Ni 0.35 max	--	0.50 max
A03540	354.0	rem	1.6-2.0	0.20 max	0.40-0.6	0.10 max	8.6-9.4	0.20 max	0.10 max	Ni 0.50 max	--	0.15 max
A03550	355.0	rem	1.0-1.5	0.6 max	0.40-0.6	0.50 max	4.5-5.5	0.25 max	0.35 max	--	0.05 max	0.15 max
A03560	356.0	rem	0.25max	0.6 max	0.20-0.40	0.35 max	4.5-5.5	0.25 max	0.35 max	Cr 0.25 max	0.05 max	0.15 max
A03570	357.0	rem	0.05max	0.15 max	0.45-0.6	0.03 max	6.5-7.5	0.20 max	0.05 max	--	0.05 max	0.15 max
A03590	359.0	rem	0.20max	0.20 max	0.50-0.7	0.10 max	8.5-9.5	0.20 max	0.10 max	--	0.05 max	0.15 max
A03800	380.0	rem	3.0-4.0	2.0 max	0.10 max	0.50 max	7.5-9.5	--	3.0 max	Sn 0.35 max, Ni 0.50 max	--	0.50 max
A05140	514.0	rem	0.15max	0.5 max	3.5-4.5	0.35 max	0.35 max	0.25 max	0.15 max	--	0.05 max	0.15 max
A05180	518.0	rem	0.25max	1.8 max	7.5-8.5	0.35 max	0.35 max	--	0.15 max	Ni 0.15max, Sn 0.15 max	--	0.25 max
A05200	520.0	rem	0.25max	0.30 max	9.5-10.6	0.15 max	0.25 max	0.25 max	0.15 max	--	0.65 max	0.15 max
A05350	535.0	rem	0.05max	0.15 max	6.2-7.5	0.10-0.25	0.15 max	--	--	Be 0.003 - 0.007, B 0.002 max, Sn 0.10 - 0.25	0.05 max	0.15 max
A07070	707.0	rem	0.20max	0.8 max	1.8-2.4	0.40-0.6	0.20 max	0.25 max	4.0-4.5	Cr 0.20 - 0.40	0.05 max	0.15 max
A12400	A240.0	rem	7.0-9.0	0.50 max	5.5-6.5	0.30-0.7	0.50 max	0.20 max	0.10 max	Ni 0.30 - 0.7	0.05 max	0.15 max
A13560	A356.0	rem	0.20max	0.20 max	0.20-0.40	0.10 max	6.5-7.5	0.20 max	0.10 max	--	0.05 max	0.15 max
A13570	A357.0	rem	0.20max	0.20 max	0.40-0.7	0.10 max	6.5-7.5	0.10-0.20	0.10 max	Be 0.04-0.07	0.05 max	0.15 max
A17120	A712.0	rem	0.35-0.65	0.50 max	0.6-0.8	0.05 max	0.15 max	0.25 max	6.0-7.0	--	0.05 max	0.15 max
A22950	B295.0	rem	4.0-5.0	1.2 max	0.05 max	0.35 max	2.0-3.0	0.25 max	0.50 max	Ni 0.35 max	--	0.35 max
A23580	B358.0	rem	0.20max	0.30 max	0.40-0.6	0.20 max	7.6-8.6	0.10-0.20	0.20 max	Be 0.10-0.30, Cr 0.20 max	0.05 max	0.15 max
A33550	C355.0	rem	1.0-1.5	0.20 max	0.40-0.6	0.10 max	4.5-5.5	0.20 max	0.10 max	--	0.05 max	0.15 max
A47120	D712.0	rem	0.25max	0.50 max	0.50-0.65	0.10 max	0.30 max	0.15-0.25	5.0-6.5	Cr 0.40-0.6	0.05 max	0.20 max
XA2201.0	XA201.0	rem	4.0-5.0	0.10	0.15-0.35	--	0.05	0.15-0.35	--	Ag 0.40-1.2	0.03	0.10
Avior (French)		rem	4.7	0.05 max	0.3	--	0.05	0.4	1.3	Ag 0.8, Ni 0.03	--	--
AlZn6Mg3 (German)		rem.	0.01	0.06	3	--	0.07	--	6	--	--	--

Based on references [1] to [6]

Table II Examples of Heat Treatments for Cast Aluminum Alloys

Alloy UNS	Designation A A	Temper	Product	Solution Heat Treatment Metal Temperature °C (°F)	Time, Hours	Aging Treatment Metal Temperature °C (°F)	Time, Hours
A02240	224.0	T7	S	538 (1000)	24	191 (375)	36
A02950	295.0	T4	S	516 (960)	12	--	--
		T6	S	516 (960)	12	154 (310)	3-6
		T7	S	516 (960)	12	154 (310)	12-24
A03190	319.0	T5	S	--	--	204 (400)	8
		T6	S	504 (940)	12	154 (310)	2-5
		T6	P	504 (940)	4-12	154 (310)	2-5
A03330	333.0	T6	P	504 (940)	6-12	154 (310)	2-5
		T7	P	504 (940)	6-12	260 (500)	4-6
A03550	355.0	T6	S	527 (980)	12	154 (310)	3-5
A03560	356.00	T6	S	538 (1000)	12	154 (310)	3-5
		T7	S	538 (1000)	12	204 (400)	3-5
A03570	357.0	T6	P	538 (1000)	8	177 (350)	6
A05200	520.0	T4	S	432 (810)	18	--	--
A07070	707.0	T7	P	532 (990)	4-8	177 (350)	4-10
A13560	A356.0	T61	P	538 (1000)	6-12	Room temperature then 154 (310)	8 min. 6
A22950	B295.0	T4	P	510 (950)	8	--	--
		T6	P	510 (950)	8	154 (310)	1-8
		T7	P	510 (950)	8	260 (500)	4-6
A33550	C355.0	T61	P	527 (980)	6-12	Room temperature then 154 (310)	8 min. 10-12
A47120	D712.0	T5	S	--	--	Room temperature or 157(315)	21 day 6-8

From references [4] [16]. S = Sand cast, P = Permanent mold cast.
Temperatures ± 6°C (± 10°F). From solution heat quench into water at
66-100°C (150-212°F).

Table 3 lists examples for typical mechanical properties
which can be achieved with cast aluminum alloys. It cannot
be overemphasized that the mechanical property data listed in
Table 3 are not guaranteed minimum values needed in design. The
data presented in Table 3 can realistically be obtained in
aluminum alloy castings, and they should serve for a general
comparison of the various alloys as well as for a comparison
with typical mechanical properties of other aluminum alloys.
However, the minimum guaranteed mechanical properties of cast
aluminum alloys can be significantly lower than those listed

Table III Typical Mechanical Properties for Commonly Used Aluminum Sand and Permanent Mold Castings

Alloy Designation UNS	A A	Temper	Tensile Strength MN/m²	(ksi)	Yield Strength MN/m²	(ksi)	Elongation	Fatigue Limit MN/m²	(ksi)
A02010	201.0	-T6	>414	(>60)	>345	(>50)	>5	--	--
		-T7	>414	(>60)	>345	(>50)	>3	--	--
A02240	224.0	-T7	421	(61)	331	(48)	4	76	(11)
A02490	249.0	-T7	469	(68)	407	(59)	6	76	(11)
A02950	295.0	-T4	>200	(>29)	>90	(>13)	6	--	--
		-T6	>221	(>32)	>138	(>20)	3	--	--
		-T7	>200	(>29)	--	--	3	--	--
A03190	319.0	-F	234	(34)	131	(19)	2.5	83	(12)
		-T5	>172	(>25)	--	--	--	--	--
		-T6	276	(40)	168	(27)	3	83	(12)
A03330	333.0	-F	234	(34)	131	(19)	2	100	(14.5)
		-T6	290	(42)	207	(30)	1.5	103	(15)
		-T7	255	(37)	193	(28)	2	83	(12)
A03540	354.0	-T6	379	(55)	283	(41)	6	117	(17)
A03550	355.0	-T6	290	(42)	168	(27)	4	69	(10)
A03560	356.0	-T6	255	(37)	168	(27)	5	90	(13)
		-T7	221	(32)	165	(24)	6	76	(11)
A03570	357.0	-T6	359	(52)	296	(43)	5	90	(13)
		-T61	>310	(>45)	>241	(>35)	>3		
A3590	359.0	-T61	324	(47)	255	(37)	7	103	(15)
		-T62	345	(50)	290	(42)	5	103	(15)
A05140	514.0	-F	172	(25)	83	(12)	9	48	(7)
A05200	520.0	-T4	331	(48)	179	(26)	16	55	(8)
A05350	535.0	-F	290	(42)	159	(23)	10	--	--
A07070	707.0	-F	290	(42)	165	(24)	14	--	--
		-T6	338	(49)	248	(36)	9.5	--	--
		-T7	365	(53)	296	(43)	6.5	--	--
A12400	A240.0	-F	228	(33)	193	(28)	1	76	(11)
A13560	A356.0	-T6	283	(41)	207	(30)	(10)	90	(13)
		-T61	283	(41)	207	(30)	10	90	(13)
A13570	A357.0	-T6	359	(52)	290	(42)	8	90	(13)
		-T61	359	(52)	290	(42)	5	103	(15)
A17120	A712.0	-F	241	(35)	172	(25)	5	62	(9)
A22950	B295.0	-T4	>228	(>33)	--	--	5	--	--
		-T6	276	(40)	179	(26)	5	69	(10)
		-T7	>228	(>33)	--	--	>3	--	--
A23580	B358.0	-T6	345	(50)	290	(42)	6	--	--
		-T62	365	(53)	317	(46)	3.5	--	--
A33550	C355.0	-T61	303	(44)	234	(34)	>3	97	(14)
A47120	D712.0	-T5	>221	(>32)	>140	(>20)	>3	--	--
--	XA201.0	-T7	496	(72)	448	(65)	6	--	--
Avior (French)		-T6	500	(73)	440	(64)	10	--	--
		-T7	490	(71)	432	(63)	10	--	--
AlZnMg3 (German)		-T6	420	(61)	420	(61)	>1	--	--

Based on references [3) - 8)]

Note these data do not represent guaranteed minimum mechanical properties for design. The data are intended only as a basis for comparing alloys and tempers.

in Table 3. Further information is available in references[3) to 8)], but it will often be necessary to negotiate specific minimum mechanical properties with the foundry since the effects of processing and the particular configuration are very strong.

The data shown in Table 3 can also serve as a basis for comparing different tempers of a given alloy. For example,

alloys are often softest in the as cast condition (-F),
hardest in the -T6 temper, and show intermediate strength
in the overaged -T7 temper, e.g. alloy 333.0.

Stress Corrosion Resistance of Various Types of Cast Al Alloys

Table 4 presents an overview over the stress corrosion
resistance of several types of cast aluminum base alloys.
Stress corrosion cracking ratings are based on service exerience
and on laboratory tests of specimens exposed to the 3.5%
sodium-chloride alternate immersion test[1)5)9)-20)].

Table IV Stress Corrosion Cracking Resistance of Cast Aluminum Alloys

Alloy Designation UNS	A A	Type of Alloy	Temper	Stress Corrosion Rating	Temper	Stress Corrosion Rating	Temper	Stress Corrosion Rating	General Corrosion Resistance
A02010	201.0	Al-Cu-Ag			-T6	D	-T7	A	4
A02240	224.0	Al-Cu			-T6	C	-T7	B	4
A02490	249.0	Al-Cu-Zn			-T6		-T7		4
A02950	295.0	Al-Cu-Si	-T4		-T6	D	-T7		4
A03190	319.0	Al-Si-Cu	as cast	B	-T5		-T6		3
A03330	333.0	Al-Si-Cu	as cast	B	-T6		-T7		3
A03540	354.0	Al-Si-Cu			-T6	B			3
A03550	355.0	Al-Si-Cu	-T51		-T6	B	-T7		3
A03560	356.0	Al-Si-Mg	-T51	A	-T6	A	-T7	A	2
A03570	357.0	Al-Si-Mg	-T6	A	-T61	A			2
A03590	359.0	Al-Si-Mg	-T6	A	-T61	B	-T62	B	2
A03800	380.0	Al-Si-Cu	as cast	B					5
A05140	514.0	Al-Mg	as cast	A					1
A05180	518.0	Al-Mg	as cast	A?					
A05200	520.0	Al-Mg	-T4	D					1
A05350	535.0	Al-Mg	as cast	A?					1
A07070	707.0	Al-Zn-Mg	as cast		-T6	D	-T7		2
A12400	A240.0	Al-Cu-Mg	as cast	B					
A13560	A356.0	Al-Si-Mg	-T51	A	-T61	A	-T7	A	2
A13570	A357.0	Al-Si-Mg	-T6	B	-T61	A	-T7	A	2
A17120	A712.0	Al-Zn-Mg-Cu	as cast	A?					4
A22950	B295.0	Al-Cu-Si	-T4		-T6	D	-T7		4
A23580	B358.0	Al-Si-Mg			-T6				2
A33550	C355.0	Al-Si-Cu			-T61	B	-T71		3
A47120	D712.0	Al-Zn-Mg	as cast	D	-T5				3
--	XA201.0	Al-Cu-Ag			-T6	C?	-T7	A?	4
Avior (French)		Al-Cu-Ag			-T6	D	-T7	A	4
AlZn6Mg3 (German)		Al-Cu-Ag	as cast	D	-T6	D			4

Based on references 1) 5) 9) - 20)

A = No known instance of SCC failure in service or laboratory tests

B = No known instance of SCC failure in service; limited failures in laboratory tests under extreme conditions

C = Ready stress corrosion failure in laboratory; SCC service failures not reported but possible

D = SCC service failures observed or anticipated

The following system of ranking was used in Table 4:

A = No known instance of SCC failures in service or
    laboratory tests.

B = No known instance of SCC failure in service; limited
    failures in laboratory tests under extreme conditions.

C = Ready stress corrosion failure in the laboratory,
    service failures not reported but possible.

D = Stress corrosion service failures observed or anticipated.

Question marks in Table 4 indicate that it is possible
the rating will be lowered as further experience is
acquired or when the casting is exposed to slightly
elevated temperatures over an extended period of time.

The general corrosion rating in Table 4 is based on the
resistance of alloys in standard salt spray tests; 1 is the
best and 5 is the poorest of the alloys listed[4][5].

Cast Aluminum-Copper Alloys of the 2XX.0 are heat treatable
and readily weldable. With small additions of silver and
magnesium 2XX.0 series alloys possess very high strength,
good ductility, and are particularly useful for high
temperature service. Examples are alloys 201.0 and XA201.0.
Also suitable for high strength and high temperature service
are the alloys 224, 249, and Avior which, in addition to
copper contain also V + Zr, or Zn, or Zn + Ag respectively.
All these alloys are inherently susceptible to stress
corrosion cracking in the peak hardness -T6 temper. Thus,
they are normally used in the -T7 temper which has a much
better stress corrosion resistance[1][7][16]. The alloy
"Avior" can serve as an example; it was developed in France
and has been marketed since the early seventies principally
for aerospace products, much like the similar American
alloys 201.0 and XA201.0. As can be seen in Table 1, the
silicon and iron contents are very small, resulting in a
high fracture toughness[1][21]. The stress corrosion resistance

of Avior in the -T73 condition is described as passing an
alternate immersion test in NaCl solution for 30 days at
80% of the yield strength. The stress corrosion crack growth
rates of "Avior" are shown in Figure 1[1]. It is apparent
that in the -T6 condition this alloy is very susceptible

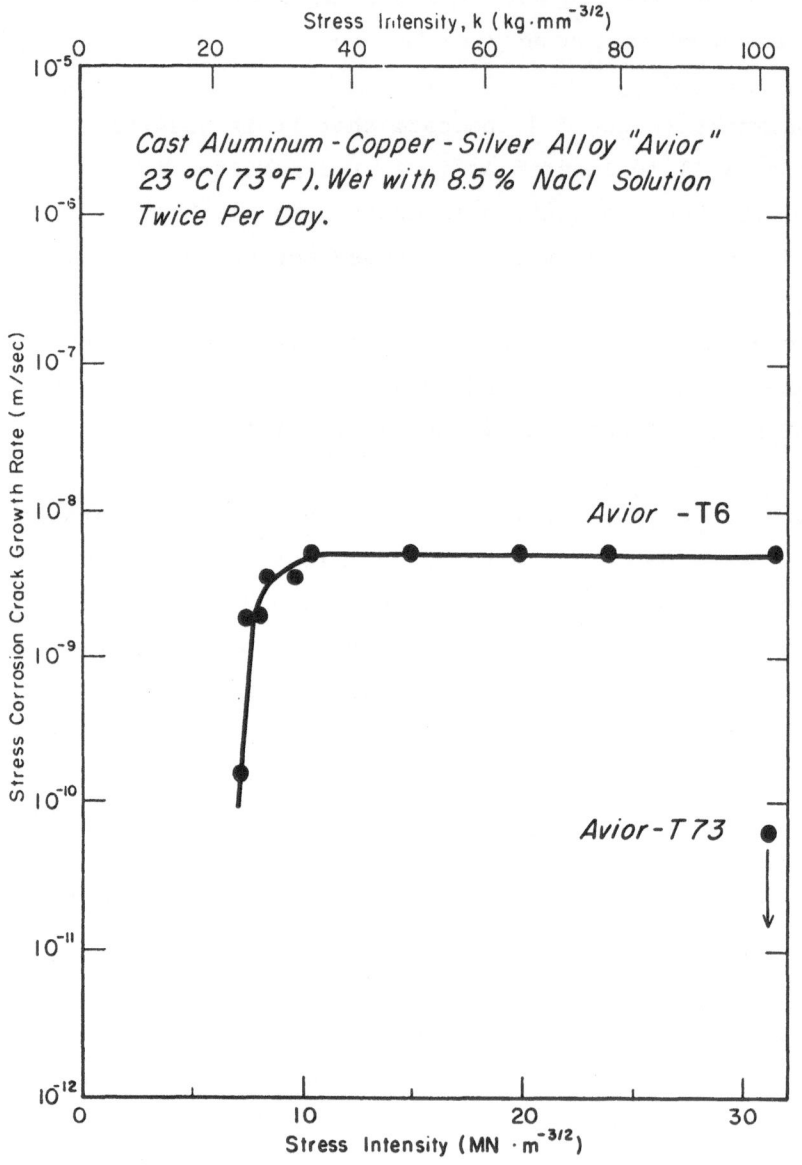

Fig. 1 Growth rate of stress corrosion cracks in a cast
high strength aluminum alloy[1]

to SCC, with a stress intensity threshold for stress
corrosion cracking below 8 MN·m$^{-3/2}$ (7ksi$\sqrt{in}$). In the
(overaged) -T7 temper, however, the alloy is completely
immune to stress corrosion cracking in 3.5% NaCl solution. A
similar behavior is exhibited by most of the 2XX.0 alloys
mentioned in this paper. Thus, premium castings of high
strength and for high temperature service are normally
offered in the -T7 condition with excellent stress corrosion
resistance[4)7)]. However, there is a scarcity of published
data on the SCC behavior of such alloys. Figure 2 shows one
of the few exceptions[16)]: it illustrates the effect of
applied stress on time to failure by SCC of the sand cast
alloy 224.0 -T7, indicating that this alloy is highly
resistant to SCC. By comparison, alloy C 355.0 -T6 is even
more resistant and alloy 295.0 -T6 is highly susceptible
to SCC. This relative performance is also reflected in
Table 4, where alloys 224.0 -T7 and 295.0 -T6 are ranked
B and D respectively.

Cast aluminum silicon alloys of the 3XX.0 series contain
added copper and/or magnesium as the principal alloying
elements. They are heat treatable, and very weldable. Because
of the high silicon content (Table 1), they are among the
easiest to cast by a variety of techniques. They have a high
resistance to corrosion, and their stress corrosion resistance
ranges from very high to excellent (Table 4). Well known
materials in the 3XX.0 series are the alloys 354.0, 355.0,
C355.0, 356.0, A356.0, A357.0, and 359.0. Stress corrosion
service failures with these alloys are not known.

Cast aluminum magnesium alloys of the 5XX.0 series are used
primarily for sand castings. They are very readily weldable.
An outstanding example is alloy 520.0. This is a relatively
high strength, heat treatable sand casting alloy with
excellent machinability and resistance to corrosion and is
very readily weldable. Unfortunately the alloy 520.0 is highly

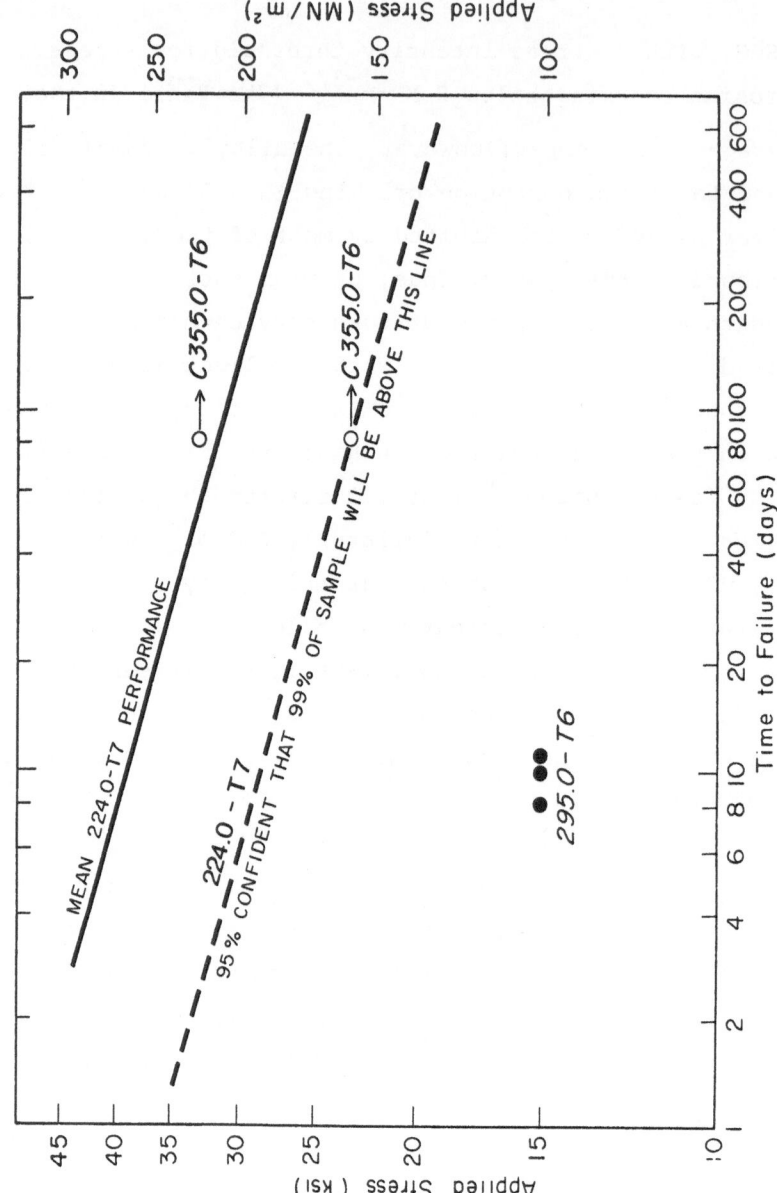

Fig. 2 Time to failure in stress corrosion tests of three
cast aluminum alloys. Smooth tensile specimens
exposed to a 3.5% sodium chloride solution by
alternate immersion[16)]

susceptible to stress corrosion cracking, and this has
caused many stress corrosion service failures in 520.0 -T4
aluminum alloy aircraft parts[12)13)]. To minimize this problem,
a "slow" quench in hot oil after solution heat treatment,
or use of an interrupted boiling water quench was recommended[5)22)]
but even so, the alloy remains susceptible to stress corrosion[22)]
Moreover, after heat treatment, subsequent heating to even
moderately elevated temperatures appreciably reduced both,
room temperature tensile properties and resistance to stress
corrosion cracking[5)]. The propensity of the high magnesium
containing cast aluminum alloys to stress corrosion cracking,
especially after some time at slightly elevated temperatures
is no surprise after all that is known about the SCC
susceptibility of wrought aluminum alloys containing more
than 3.5% magnesium[23)].

Cast aluminum zinc alloys of the 7XX.0 series generally also
contain smaller amounts of magnesium. Examples are 707.0,
A712.0, and D712.0. The tensile properties of these alloys
in the as cast condition increase rapidly during the first
weeks of room temperature aging, due to precipitation
hardening. Thus, the alloys develop a high strength without
solution heat treatment. At the same time, the stress corrosion
resistance deteriorates steadily[10)]. Heat treatments of the
-T6 and -T7 types may be applied to alloy 707.0. The eutectic
melting points of alloys in this group are high, which makes
them suitable for castings that are to be assembled by brazing.

Alloys of the 7XX.0 group are not commercially weldable
and are not recommended for service at elevated temperature.
Their corrosion resistance is not as good as that of the
5XX.0 group.
The high SCC susceptibility of alloy D712.0 and other cast
Al-Zn alloys is well known and has caused many stress corrosion
service failures[11)15)17)]. Thus, other alloys are now chosen
for service under conditions where stress corrosion cracking

in D712.0 is a consideration[17] . Exposure to salt water does
not appear to be necessary for stress corrosion cracking to
occur in cast aluminum-zinc alloys, since SCC service
failures have been observed upon exposure to an industrial
atmosphere[11] .

Stress Corrosion Fracture Path in Cast Aluminum Alloys

Like in wrought aluminum alloys, the path of stress corrosion
cracks in cast aluminum alloys is almost exclusively intergranular, i.e. the cracks follow the grain boundaries. This
is illustrated in Figure 3[18] and Figure 4[17] for an Al-Cu-Ag
alloy and an Al-Zn-Mg alloy.

In contrast to most wrought aluminum alloys, however, the
cast aluminum alloys have aquiaxed grains and thus, in SCC
susceptible material there is no "relatively safe" orientation
like the longitudinal direction in wrought alloys. Therefore,
stress corrosion susceptible cast aluminum alloys will be
equally susceptible (or equally resistant) to stress corrosion
cracking in all directions. While this is true for the
growth of stress corrosion cracks, the initiation of such
cracks depends on the surface condition and this, of course
can be very differnt in different locations of a casting.

Comparison of SCC of cast and wrought Al alloys

It has been assumed that generally, cast aluminum alloys
are more resistant to stress corrosion cracking than wrought
alloys[11] . However, in view of the service failures encountered
and in view of newer laboratory data there does not seem to
be a significant inherent difference in the SCC susceptibility
of the two classes of alloys. The fact that the number of
SCC service failures encountered with aluminum alloy castings
is moderate has been attributed to generally lower tensile
stresses and the protection provided by the as cast surface
condition. However, once machining has removed the protective

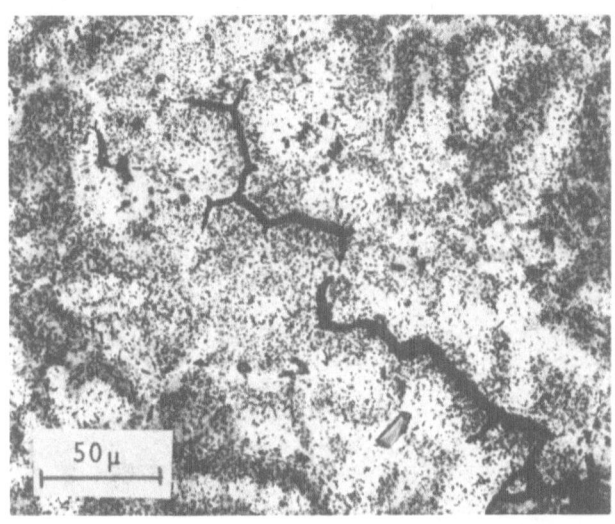

Fig. 3 Intergranular stress corrosion crack in a cast
Al-Cu-Ag alloy which was exposed to NaCl solution[18]

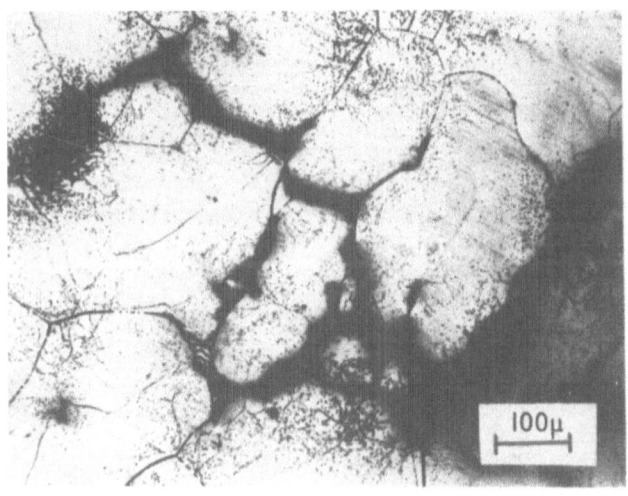

Fig. 4 Intergranular stress corrosion crack in a cast
Al-Zn-Mg alloy, exposed to NaCl solution[17]

surface, SCC susceptible cast alloys can develop stress corrosion cracks in all orientations, due to the equiaxed grain structure.

The growth rates of stress corrosion cracks in wrought and cast aluminum alloys are very similar. This is shown in Figure 5, where the growth rate of stress corrosion cracks in the cast alloy Avior (of the Al-Cu-Ag type) is plotted

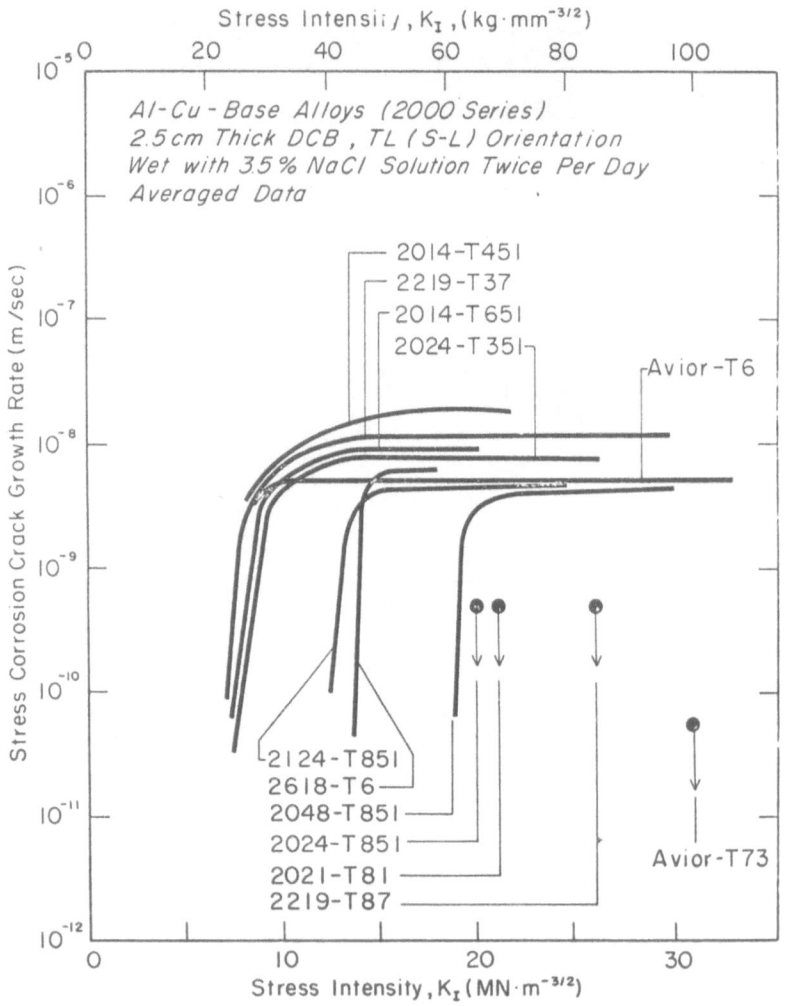

Fig. 5   Stress corrosion crack growth rates in copper-containing wrought and cast aluminum base alloys[1]

versus the applied stress intensity, together with similar data from many wrought aluminum alloys of the Al-Cu-Mg type[1]. In the -T7 condition, however, the cast alloy has excellent stress corrosion resistance even when stressed close to K_{Ic} and exposed to conditions that readily cause cracking in most wrought alloys of the 2XXX series.

A comparison of cast and wrought alloys is also possible on the basis of time-to-failure curves. Figure 6 shows that a cast Al-6%Zn-3%Mg alloy has a stress corrosion threshold stress of about 50 MN/m^2, corresponding to about 7 ksi[10]. This is the same stress corrosion threshold stress a very

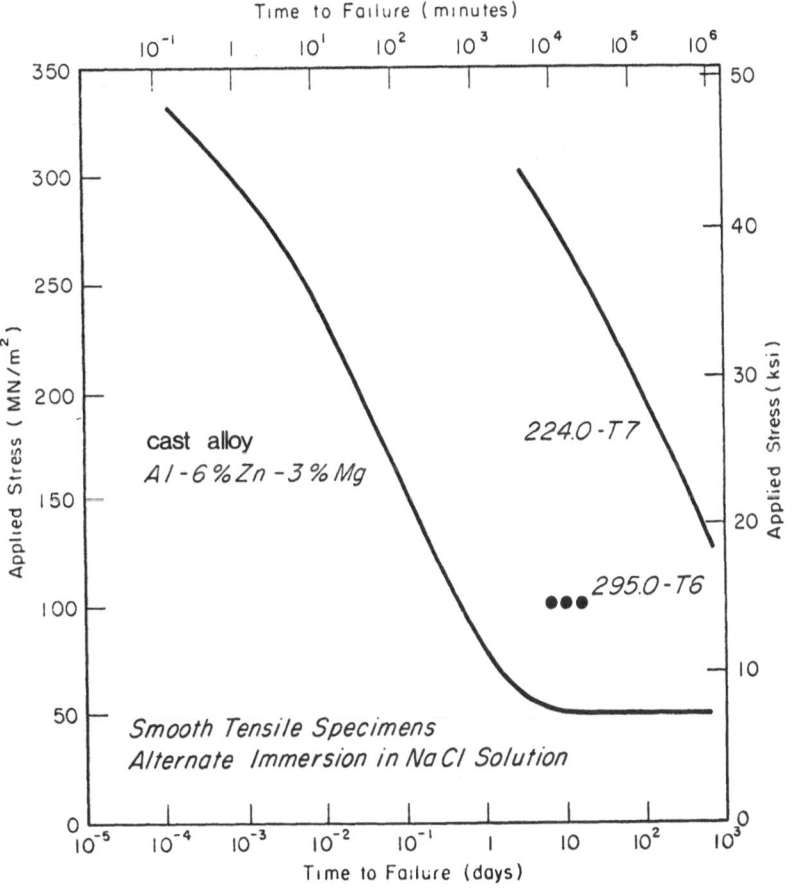

Fig. 6   Times to failure as functions of applied stress for three cast aluminum alloys[10][16]

SCC susceptible wrought alloy of the 7XXX (Al-Zn-Mg-Cu) series would have under similar conditions when stressed in the short transverse direction[1]. Such comparisons confirm that cast aluminum alloys are not inherently more SCC resistant than wrought alloys.

Figure 6 illustrates the moderate slope of the stress versus time to failure curve which is characteristic for cast aluminum alloys. For the more SCC resistant alloys like 224.0 -T7, extraordinary long testing times would be required to determine SCC threshold stress levels. This may be one reason for the dearth of such data in the published literature.

Summary of Stress Corrosion Cracking of Cast Aluminum Alloys

The stress corrosion performance of aluminum alloy castings has much in common with that of wrought alloys:
- High contents of the alloying elements zinc, magnesium, and copper, separately or together can render cast aluminum alloys susceptible to stress corrosion cracking.
- Aluminum castings with silicon as the major alloying element, and overaged copper containing alloys can have excellent resistance to stress corrosion.
- Residual tensile stresses and installation stresses are the cause of most SCC service failures; service stresses seldom lead to failure by stress corrosion.
- Moist atmospheres are aggressive enough to cause stress corrosion cracking in some high strength aluminum castings. Salt water accelerates stress corrosion failures.
- Improved alloys and heat treatments are now available which allow a combination of high strength and high resistance to stress corrosion cracking.

References

1) Markus O. Speidel, "Stress Corrosion Cracking of Aluminum Alloys", Met. Trans., Vol. 6A (1975) p. 631-651

2) ASTM DS-56, "Unified Numbering System for Metals and Alloys", ASTM Race Street, Philadelphia, USA, 1975

3) "Aluminum Standards and Data", the Aluminum Association, New York, 1974

4) "Standards for aluminum sand and permanent mold castings", the Aluminum Association, New York, 1974

5) "Aluminum", Kent R. Van Horn, ed., Vol. I., ASM, Metals Park, Ohio, 1971

6) D. Altenpohl, "Aluminium und Aluminiumlegierungen", Springer Verlag, Heidelberg, 1965

7) "Premium Castings", Premium Castings Division of ALCOA Corona, California, 1972

8) "ALCOA Aluminum Design Data", Aluminum Company of America, Pittsburgh, USA

9) Heinz Borchers und Erich Tenckhoff, "On the Influence of the Surface Structure on the Susceptibility to SCC of Al-Zn-Mg Cast Alloys", Z. Metallkunde Vol. 59 (1968) p. 58 - 62

10) H. Borchers and E. Tenckhoff, "Ueber das Spannungsriss-korrosionsverhalten von Al-Zn-Mg Gusswerkstoff", Werkstoffe und Korrosion 20 (1969) p. 319-323

11) Hugh L. Logan, "The Stress Corrosion of Metals", John Wiley, New York, 1966

12) Fred M. Reinhart, "The Effect of Heat Treatment on the Susceptibility of Sand Cast Aluminum Alloy 220 to Stress Corrosion Cracking, Corrosion Vol. 13, (1957), p. 17-18

13) Fred M. Reinhart and William F. Gerhold, "Stress Corrosion of High Strength Cast Aluminum Alloys", Corrosion, Vol. 18 (1962), p. 158-162

14) D.O. Sprowls and Herbert C. Rutemiller, "Susceptibility of Aluminum Alloys to Stress Corrosion", Materials Protection (1963) p. 63-65

15) E.H. Spuhler, C.L. Burton, and J.A. Dickson, "Avoiding Stress Corrosion Cracking in High Strength Aluminum Alloy Structures", ALCOA Green Letter, Aluminum Company of America, Pittsburgh, 1970

16) John S. Lengel, Private communication, ALCOA, 1975

17) H.P. van Leeuwen, Private communication, NLR, Amsterdam, 1975

18) Markus O. Speidel, unpublished data, 1975

19) MIL - Handbook - 5B, U.S. Government Printing Office, Washington, D.C., 1971

20) Aerospace Structural Metals Handbook Vol. 2, Mechanical Properties Data Center, Traverse City, Michigan, 1974

21) Markus O. Speidel, "Development of High Strength Aluminum Alloys", to be published in "Aluminum", 1976

22) Discussion to ref. 12), Corrosion Vol. 13 (1957)p.420

23) M.O. Speidel and M.V. Hyatt "Stress Corrosion Cracking of High Strength Aluminum Alloys" <u>in</u> Advances in Corrosion Science and Technology, Vol.2, Plenum Press, New York, 1972

STRESS CORROSION AND CORROSION FATIGUE

CRACK GROWTH IN ALUMINUM ALLOYS

Markus O. Speidel[*]

Department of Metallurgical Engineering

Ohio State University, Columbus, Ohio USA

Summary

The growth of subcritical cracks in aluminum alloys exposed
to environments of different aggressiveness is analyzed.
It is shown that cyclic loads can cause three different
modes of crack extension:

- fatigue
- corrosion fatigue   and
- stress corrosion under cyclic loads.

In commercial aluminum alloys, the three modes of crack
extension often differ in their fracture path. Fatigue
crack growth is generally transgranular and stress corrosion
cracking is generally intergranular; corrosion fatigue can
result in either trans- or intergranular cracking.

---

[*] On leave of absence from Brown Boveri Research Center,
CH-5401 Baden, Switzerland

Three groups of parameters influence the growth rate of
subcritical cracks:

- mechanical
- metallurgical   and
- environmental   parameters

Of the mechanical parameters, the cyclic stress intensity
range and the cyclic load frequency are discussed in detail.

Introduction

Environment-assisted subcritical crack growth under sustained
load (stress corrosion cracking, SCC) and under cyclic loads
(corrosion fatigue, CF) are significant failure mechanisms in
aluminum alloys, just as in many other materials. This paper
reviews the mechanical, metallurgical, and environmental effects
on the growth of stress corrosion and corrosion fatigue cracks.

The initiation and the propagation of environment - assisted
subcritical cracks are different, but perhaps equally important,
aspects of stress corrosion cracking and corrosion fatigue.
The present paper is limited to the growth aspect of stress
corrosion cracks and corrosion fatigue cracks, and this
appears well justified in the light of present design
philosophies and specifications for high-performance structures.
However, it might be useful to keep in mind that consideration
of crack growth alone could lead to over-conservative design
in situations where crack initiation is the dominating
factor in the overall time to failure.

The distinction between crack growth and crack initiation
has been very helpful in elucidating the more fundamental
and theoretical aspects of stress corrosion cracking and
corrosion fatigue. The limited theoretical understanding

of stress corrosion crack growth in aluminum alloys has been
summarized at the Nato Science Committee Research Evaluation
Conference in Portugal 1971[1], and the present paper assumes
that the reader has access to that reference[1]. Recent
reviews have also included engineering aspects and initiation
of stress corrosion cracks from smooth specimens[2,3,4] as
well as corrosion fatigue crack growth[5,6] of aluminum base
alloys.

It is shown that, at a given stress intensity range, "true
corrosion fatigue" and "stress corrosion under cyclic loads"
differ strongly in their frequency response.

Of the metallurgical parameters, fracture toughness, the
orientation of grain boundaries with respect to cracks,
and the susceptibility to stress corrosion cracking, each
strongly influence the growth rate of subcritical cracks
under cyclic loads. In the absence of stress corrosion
cracking, modern high strength aluminum alloys do not
differ greatly in corrosion fatigue crack growth.

Of the environmental parameters, aqueous solutions are
discussed in detail. It is shown that certain additions
to water accelerate crack growth, others inhibit, and still
others have no effect at all. The really damaging species
are chloride, bromide, and iodide ions, just as in stress
corrosion cracking under sustained loads. Cathodic protection
and inhibitors such as fluoride, chromates and nitrate can
retard corrosion fatigue crack growth. This is compared to
similar effects on stress corrosion crack growth.

Since the NATO Portugal Conference 1971, significant progress
has been made in the theoretical understanding of the
mechanisms that are involved in stress corrosion cracking of
aluminum alloys and in our general understanding of stress
corrosion crack growth[4,7-21]. However, this appears not
to be the right time for an updated review of the theories
of stress corrosion cracking since there is still no

generally accepted concept emerging. Instead, we shall
concentrate in this paper on the effect cyclic loads have
on stress corrosion cracking, and on the effect solutions
normally causing stress corrosion cracking have on corrosion
fatigue crack growth.

In order to provide the necessary background, we first
review briefly recently developed aluminum alloys, their
resistance to stress corrosion cracking, their fracture
toughness, and their fatigue resistance. Then we discuss
in detail corrosion fatigue and stress corrosion cracking
under cyclic loads.

Fracture Mechanics

Stress corrosion crack growth can be considered as environment-
assisted subcritical crack growth under sustained loads.
This has been discussed in detail in reference[1]. Since that
time extensive work has been done in this and related areas
of fracture studies. Among other results, two analogies
emerged between subcritical crack growth, namely the effect
of stress intensity on the crack velocity and the effect
of the loading rate on the crack toughness.

The effect of stress intensity on the crack velocity is
shown schematically in Fig. 1. Let us first consider a
precracked specimen under rising load in an inert environment.
As the load increases, so does the crack tip stress intensity
because of the general relation

$$K = \sigma \sqrt{a} \cdot f(a) \tag{1}$$

where K is the stress intensity factor, σ the nominal stress,
a   the length of the precrack and f(a) a correction factor
which also depends on the specimen geometry. As the stress

intensity rises, it eventually reaches a critical value, K_{Ic}, at which a preexisting crack starts to propagate fast. As can be seen in Fig 1, the velocity of such a "super-critical" crack increases sharply above K_{Ic}, but the maximum growth rate of such a purely mechanical crack is limited to several thousand meters per second, corresponding to the velocity of elastic waves in the solid. Thus a plateau is reached at high stress intensities in the crack velocity versus stress intensity curve (v-K curve), Fig. 1.

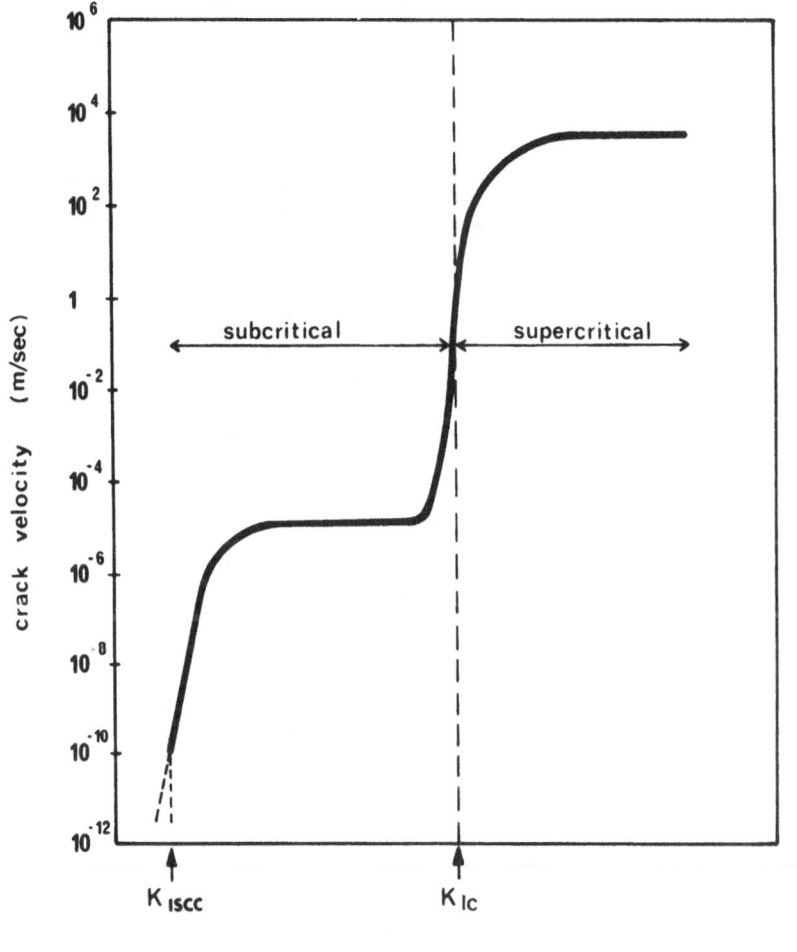

Fig. 1   Schematic representation of the effect of stress intensity on the velocity of subcritical (stress corrosion-) and supercritical (mechanical-) cracks

In inert environments, there should be no significant crack growth below K_{Ic}. However, in environments conducive to stress corrosion cracking, environment-induced subcritical crack growth is observed under sustained loads with stress intensities between K_{Ic} and K_{Iscc}, as discussed in reference[1]. As shown in Fig. 1, both, subcritical and supercritical cracking exhibit a stress intensity dependent crack velocity at relatively small stress intensities, i.e. just above K_{Iscc}, respectively K_{Ic}. At intermediate stress intensities, the crack velocity is independent of the stress intensity, and the v - K curve shows plateaus for both subcritical and supercritical crack growth. It is widely agreed that a transport mechanism is limiting the growth rate of cracks in the plateau region of the subcritical part of the v-K curve, while the limited speed of propagation of elastic waves is responsible for the plateau in the supercritical part of the v - K curve.

Another analogy between subcritical crack growth (stress corrosion) and supercritical crack growth (dynamic overload failure) is illustrated in Fig. 2. Consider a precracked specimen being loaded in an environment which is conducive to stress corrosion cracking. At a reasonably fast loading rate, say, 1 to 10 $MN \cdot m^{-3/2} \cdot sec^{-1}$, the effect of the environment on the fracture toughness K_{Ic} will be negligible, and the toughness is measured as the material constant K_{Ic}. As the loading rate increases, the toughness passes through a minimum, labeled K_{Id} in Fig. 2, for dynamic fracture toughness. Adiabatic effects will eventually force an increase in the dynamic fracture toughness at even higher loading rates. (For aluminum alloys, this remains to be investigated and to be observed)[22].

As shown in Fig. 2, a similar toughness trough may be expected for small loading rates, leading to stress corrosion cracking.

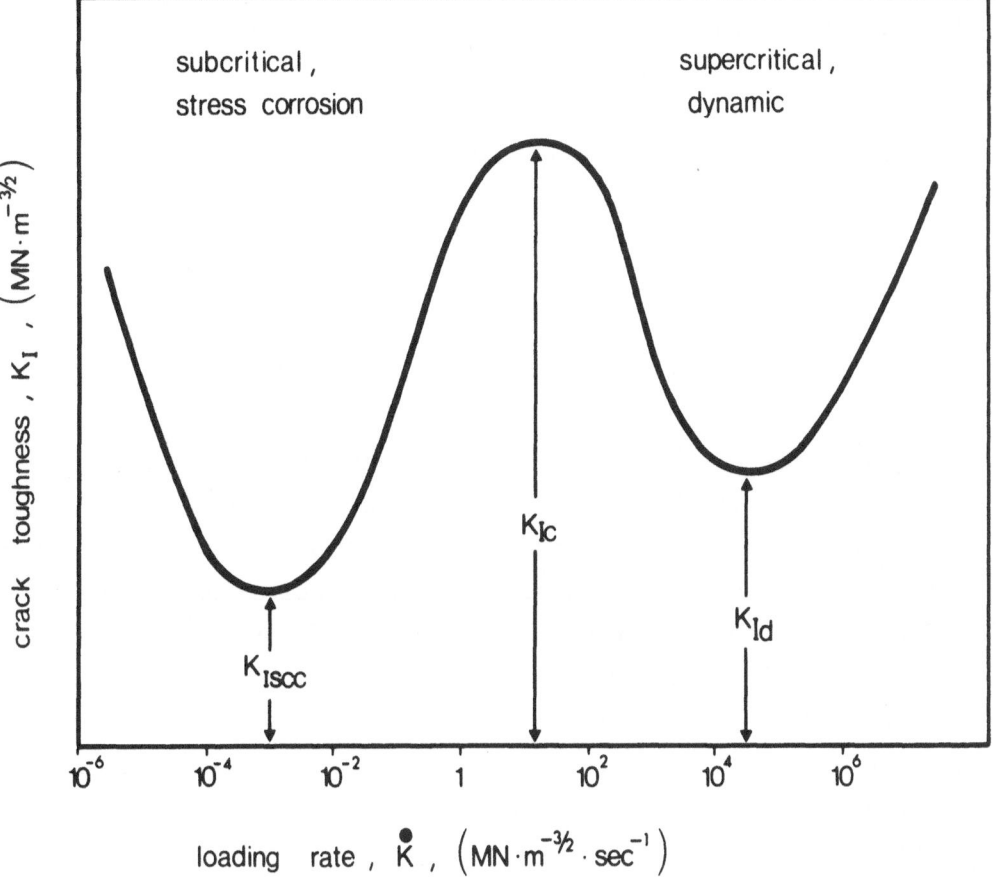

Fig. 2 Schematic representation of the effect of the
loading rate K̇ on the toughness of precracked
specimens in environments which are conducive
to stress corrosion cracking

The decrease of the crack resistance with decreasing loading
rates in aggressive environments has indeed been observed
for certain material-environment combinations which are
susceptible to stress corrosion cracking[23]. The reason for
the decrease of toughness from K_{Ic} to K_{Iscc} with decreasing
loading rate is seen in the stress corrosion crack growth
more and more outrunning the accumulation of stress and strain
which would otherwise lead to mechanical fracture at high
loading rates. The final increase of toughness with very

low loading rates as shown in Fig. 2 has not yet been
investigated with precracked specimens. It is therefore
just a hypothesis, based on analogous results which were
obtained with smooth stress corrosion specimens at very
low strain rates[24]. The reason for this increase is thought
to be the repair of damaged surface films in the highly
strained crack tip region outrunning the film damage
mechanisms.

Stress corrosion cracking is not the only kind of environment-
assisted subcritical crack growth; it is just the one aspect
of it which is normally observed under sustained loads.
In reality, cyclic loads very often occur in structures
where stress corrosion may be a problem. Thus, corrosion
fatigue, i.e. environment-assisted subcritical crack growth
under cyclic loads has received some attention in recent
years. The relation between stress corrosion and corrosion
fatigue is one of the main topics of this paper. Fig. 3
shows a schematic comparison of the kind of results typically
observed in stress corrosion and in corrosion fatigue crack
growth rate studies. The crack tip stress intensity has been
found to correlate well the mechanical driving force for both
kinds of subcritical crack growth. Time-dependent stress
corrosion crack growth is observed between K_{Iscc} and K_{Ic}.
Cycle-dependent corrosion fatigue crack growth is observed
between ΔK_o and K_{Ic}, where ΔK_o is an experimentally
observed lower limit of the cyclic stress intensity range
below which fatigue crack growth is not observed.

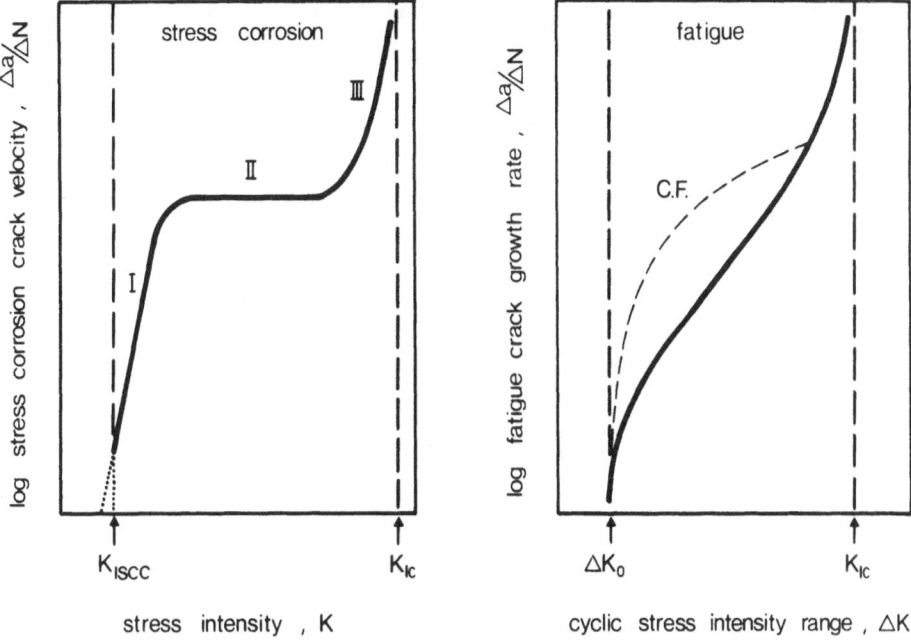

Fig. 3   Two kinds of environment-assisted subcritical
          crack growth: stress corrosion cracking (SCC) and
          corrosion fatigue (CF)

Fig. 4 presents a summary of the more conventional and the
fracture mechanics criteria which identify the limits of
stress and stress intensity below which no overload failure,
or no stress corrosion, or no fatigue is observed. For smooth

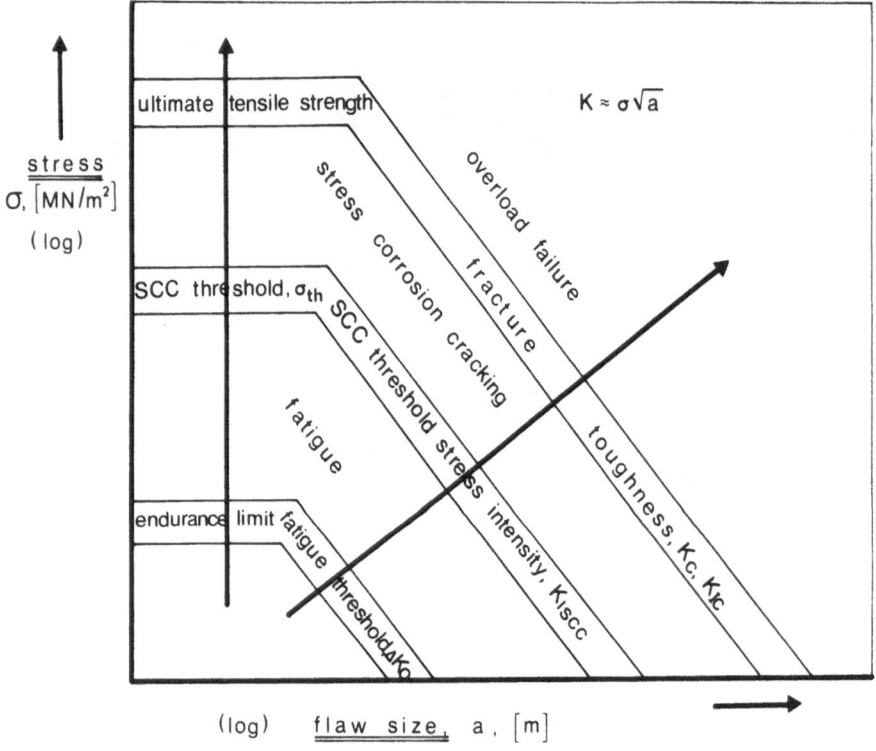

Fig. 4   Flaw size and materials properties which limit
          the resistance to fracture under monotonic loads,
          in aggressive environments, and under cyclic loads

specimens or components whith extremely small flaws, the
ultimate tensile strength, the stress corrosion threshold
stress, and the (corrosion) fatigue endurance limit are
applicable. For precracked specimens and components with
significant flaws, the fracture toughness K_{Ic}, the SCC
threshold stress intensity K_{Iscc}, and the fatigue threshold
stress intensity ΔK_o define the limits of safe loads and
failure. The arrows in Fig. 4 indicate the goals of alloy
development, that is to increase those limiting material
properties to ever higher values.

New aluminum alloys, higher toughness

The stress corrosion results presented in reference[1] have
been obtained mainly with the American high strength aluminum
alloys which were widely used in the sixties and early
seventies : alloys 7079, 7075, 7178, 7039, and 2024. The
composition of these alloys is given in Table I, together
with the composition of newly developed high strength aluminum
alloys. The newer alloys contain much less iron and silicon
and this results in greatly improved fracture toughness.
In this way alloys 2124 and 2048 have been developed from
2024; 2419 from 2219; 2214 from 2014. 7475 and 7175 are high
toughness versions of 7075, and the new 7XXX series alloys,
7050 and 7149 also achieve superior toughness due to lower
contents in iron and silicon.

Fig. 5 and Fig. 6 show the effect of the reduced iron and
silicon content on the plane strain fracture toughness of
Al-Cu-Mg and Al-Zn-Mg-Cu alloys. These dramatic improvements
are mostly due to the elimination of coarse insoluble
particles of intermetallic phases. Such particles, $\approx 1 \mu m$
to $\approx 10 \mu m$ in size, form during casting or subsequent
processing as either virtually insoluble Al_7Cu_2Fe, Mg_2Si,
or $(Fe, Mn)Al_6$, or relatively soluble $CuAl_2$, or $CuAl_2Mg$.
These constituent particles break easily when stressed at

Table I. Designation and Composition Limits of some High-strength Aluminum Alloys

Alloy Designation	Al	Cr	Cu	Fe	Mg	Mn	Si	Ti	Zn	other
					Chemical composition, weight -%					
2014	rem	0.10	3.9-5.0	0.7	0.20-0.8	0.40-1.2	0.50-1.2	0.15	0.25	
2024	rem	0.10	3.8-4.9	0.50	1.2-1.8	0.30-0.9	0.50	-	0.25	
2048	rem	-	2.8-3.8	0.20	1.2-1.8	0.20-0.6	0.15	0.10	0.25	
2124	rem	0.10	3.8-4.9	0.30	1.2-1.8	0.3-0.9	0.20	0.15	0.25	V0.05-0.15
2219	rem	-	5.8-6.8	0.30	0.02	0.20-0.40	0.20	0.02-0.10	0.10	Zr0.10-0.25
2618	rem	-	1.9-2.7	0.9-1.3	1.3-1.8	-	0.25	0.04-0.10	-	Ni0.9-1.2
Avior	rem	-	≈4.7	≈0.05	≈0.3	-	-	≈0.4	≈1.3	Ag≈0.8
7039	rem	0.15-0.25	0.10 max	0.40	2.3-3.3	0.10-0.40	0.30	0.10	3.5-4.5	
7049	rem	0.10-0.22	1.2-1.9	0.35	2.0-2.9	0.20 max	0.25	0.10	7.2-8.2	
7050	rem	0.04	2.0-2.6	0.15	1.9-2.6	0.10 max	0.12	0.06	5.7-6.7	Zr0.08-0.19
7075	rem	0.18-0.35	1.2-2.0	0.50	2.1-2.9	0.30 max	0.40	0.20	5.1-6.1	
7079	rem	0.10-0.25	0.40-0.8	0.40	2.9-3.7	0.10-0.30	0.30	0.10	3.8-4.8	
7175	rem	0.18-0.30	1.2-2.0	0.20	2.1-2.9	0.10	0.15	0.10	5.1-6.1	
7178	rem	0.18-0.35	1.6-2.4	0.50	2.4-3.1	0.30	0.40	0.20	6.3-7.3	
7475	rem	0.18-0.25	1.2-1.9	0.12	1.9-2.6	0.06	0.10	0.06	5.2-6.2	

Note: 1) generally other elements 0.05 max each and 0.15 max total

2) where no range is indicated, values are maximum contents

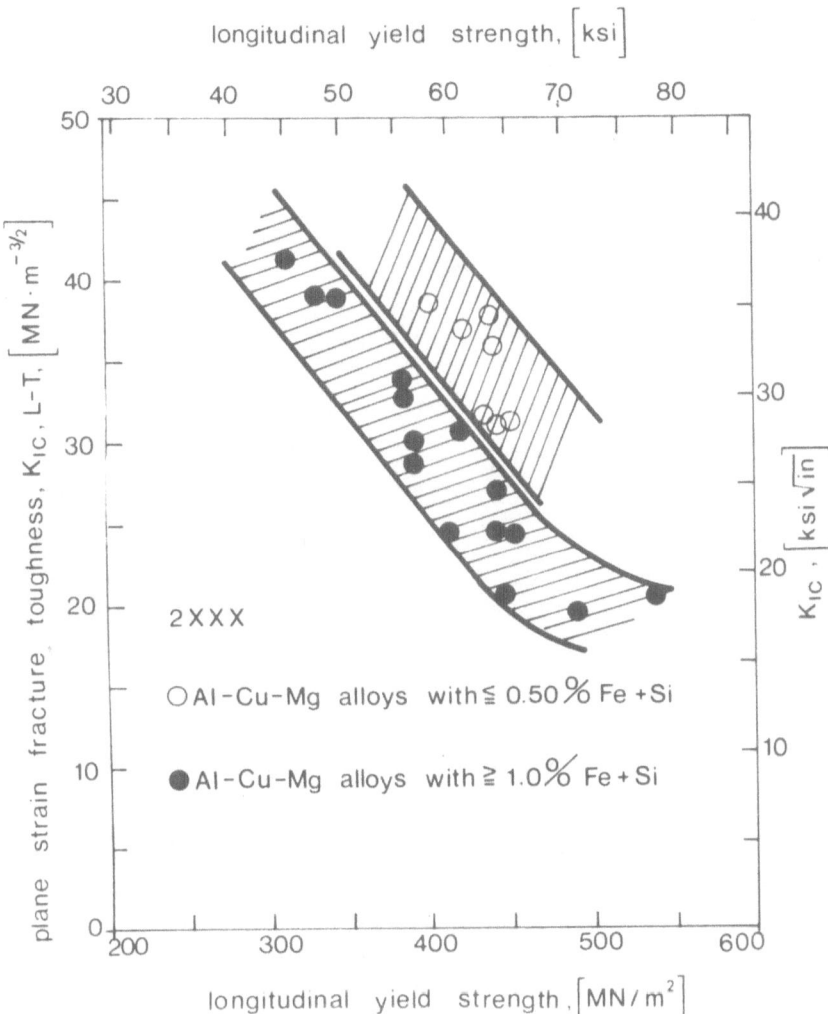

Fig. 5   Plane strain fracture toughness of old and new
          Al-Cu-Mg alloys

low stress levels, and voids develop at the cracked particles.
The first three kinds can be eliminated by eliminating Fe
and Si, the others can be reduced by reducing the copper
content, e.g. in alloy 2048 versus 2024, or they are
reduced in size by special processing of some of the new
aluminum alloys.

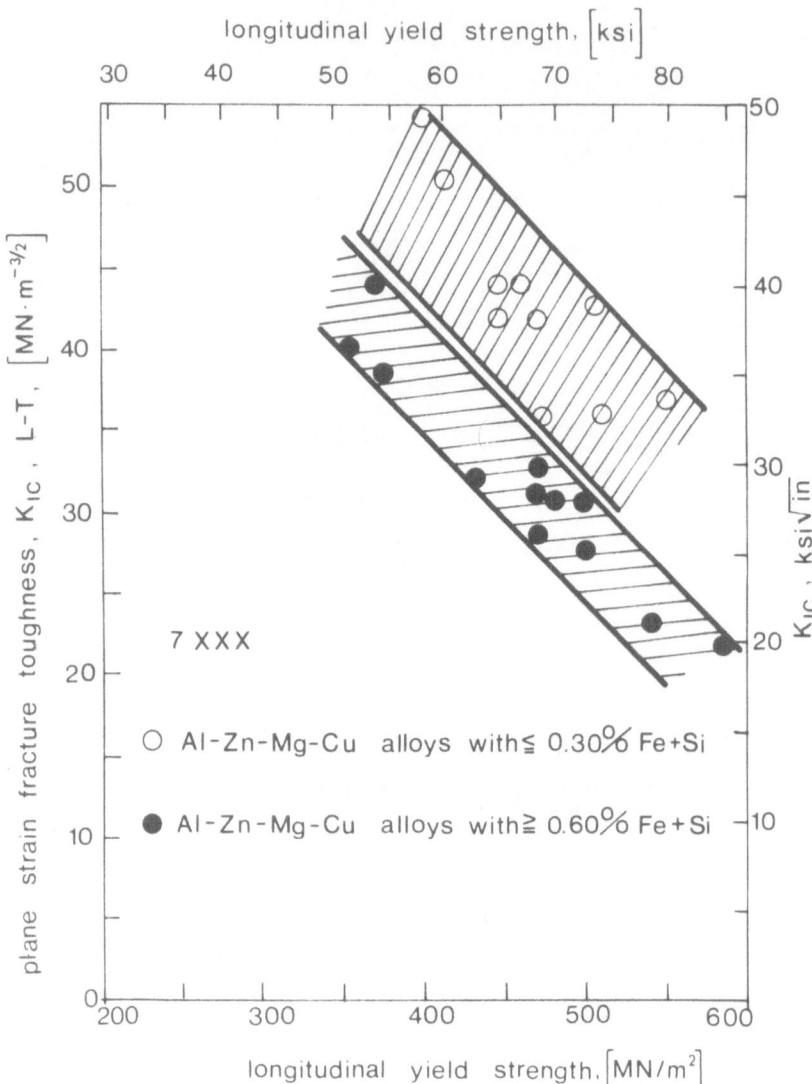

Fig. 6   Plane strain fracture toughness of old and new
         Al-Zn-Mg-Cu alloys

Fatigue crack growth in air

The growth rates of fatigue cracks in high-strength aluminum
alloys in air are shown in Fig. 7 as functions of the
applied cyclic stress intensity range, ΔK. The individual

curves shown here each represent the average of many tests;
they are the centerlines of experimental scatterbands, and
they have been collected from a large number of recent
published and unpublished reports, including the authors own.
Many different fracture mechanics specimen types (CNP,
DCB, CT) were used to obtain the test results presented in
Fig. 7, but it is important to note that all cracks were of

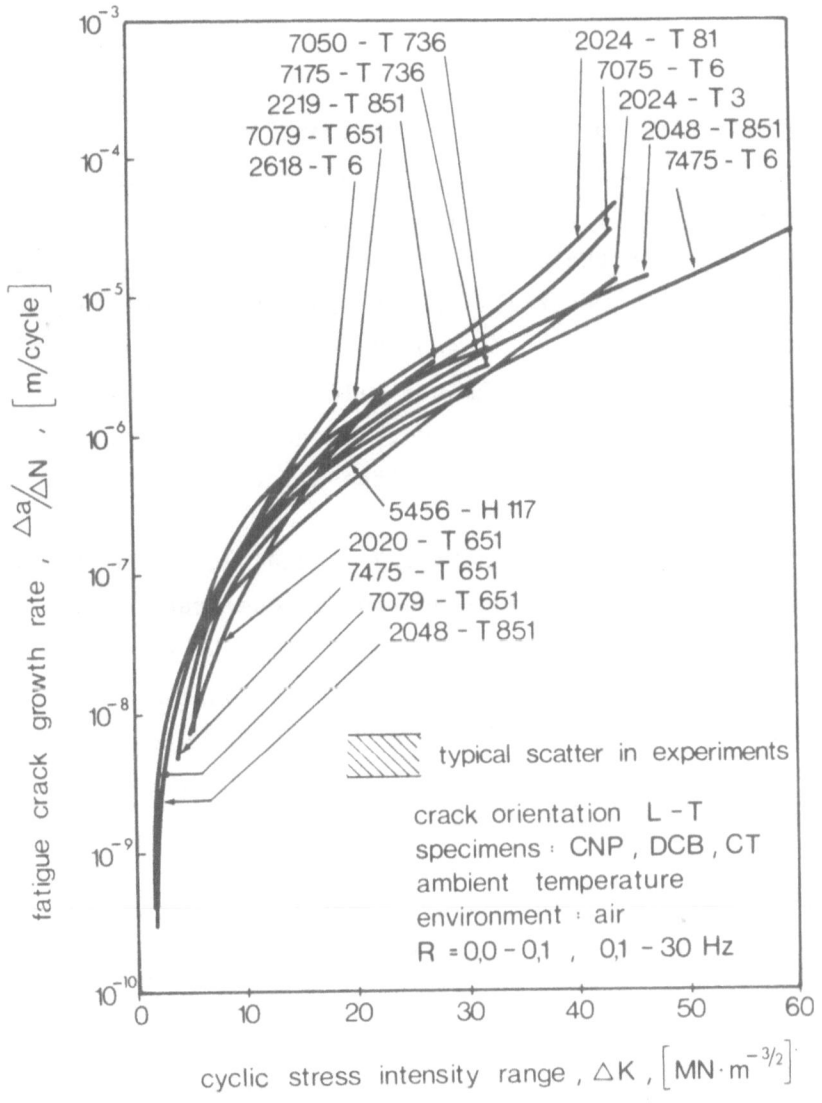

Fig. 7   Average growth rates of fatigue cracks in high
strength aluminum alloys in air

the L-T orientation, i.e. the alloys have been stressed in the longitudinal orientation. (Crack orientations are designed with a two letter system. The first letter indicates the direction of applied stress and the second letter indicates the direction of crack propagation. L = longitudinal, T = transverse, S = short transverse). For all tests reported in Fig. 7 sinusoidal load wave shapes were used with a ratio of minimum to maximum stress intensity, R, of less than 0, 1.

Two trends are apparent in Fig. 7. First, at low and intermediate stress intensity ranges, the fatigue crack growth rates of all aluminum alloys are similar. The cyclic stress intensity range, ΔK, has obviously a much more dominant influence than alloy composition and microstructure. This is consistent with the observation that fatigue crack growth rates of different materials are mainly determined by their modulus of elasticity[5)6)25)] and the fact that the modulus of elasticity of all the aluminum alloys investigated was within the limits of 71 ± 2 GN/m^2, independent of the microstructure and composition of the alloys.

The second trend observed in Fig. 7 is that aluminum alloys tend to exhibit a marked increase in fatigue crack growth rate as the cyclic stress intensity range, ΔK, approaches the fracture toughness, K_c. This has long been known and expressed mathematicalli in Formans law[26)]. As a result, the fatigue crack growth rates at high stress intensity ranges rank in the same manner as fracture toughness[27)]. This is seen in Fig. 7, where at cyclic stress intensities higher than, say, 40 MN·m$^{-3/2}$ the newer, tougher alloys 7475-T6 and 2048-T851 exhibit a lower fatigue crack growth rate than the older, not so tough alloys 7075-T6 and 2024-T81. In this way, the development of the newer, tougher alloys has helped to reduce fatigue crack growth rates; but, as seen in Fig. 7, this is pronounced only at relatively high ΔK values. At intermediate and low ΔK, the difference in fatigue crack growth rate between older alloys and such new develop-

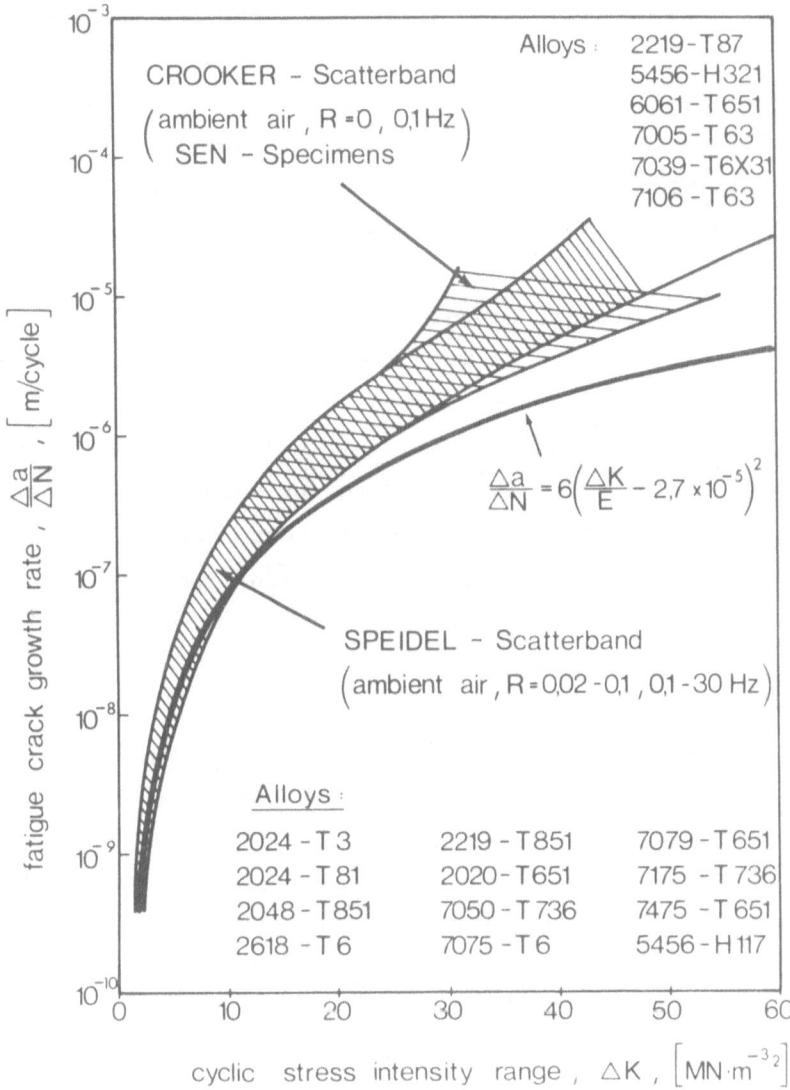

Fig. 8 Summary of measured growth rates of fatigue cracks
in aluminum alloys in air

ments as 2048 and 7475 does not significantly exceed the
typical experimental scatter.

The observation that the growth rates of fatigue cracks in
different aluminum alloys do not differ significantly at

intermediate and low stress intensity ranges is supported
by the results of Crooker[27] which have been incorporated in
Fig. 8. This figure shows the scatterband within which all
curves of Fig. 7 lie. Superimposed is a scatterband within
which all the fatigue test results of Crooker lie. Those
results had been obtained with yet another specimen type
(single edge notched specimen, = SEN) and with six aluminum
alloys different from those listed in Fig. 7. At intermediate
cyclic stress intensity ranges, the scatterband encompassing
Crookers data is identical with the scatterband encopassing
the data from Fig. 7, thus proving the point, that the
growth rates of fatigue cracks in different aluminum alloys
are all very similar at intermediate ΔK values. At higher
ΔK values, Crookers scatterband is wider, reflecting the
fact that he investigated alloys of a wider range of fracture
toughness than those shown in Fig. 7.

Fig. 8, then shows that the growth rates of fatigue cracks
in 18 different aluminum alloys fall within a well defined
scatterband at intermediate and low ΔK values. This is all
the more remarkable as the test results come from ten
different laboratories and four different types of specimens
(DNP, DCB, CT, SEN) had been used in the testing. We can
use this scatterband as a baseline for comparisons of
fatigue crack growth rates in air and different environments.

A rationalization of this observation is illustrated in
Fig. 9. A semiempirical equation for the fatigue crack growth
rate in dry air[28] predicts a common fatigue threshold stress
intensity, ΔK_o, for all aluminum alloys[5], and a sharp
increase in the crack growth rate as ΔK approaches K_c.
This equation gives reasonable predictions for fatigue
in steels[28], and Fig. 10 shows that is also works
satisfactorily for aluminum alloys of widely different
toughness.

This is apparent from the good fit of the experimental

data (squares, circles, and triangles) and the prediction
(continuous black lines) based on the equation

$$\frac{\Delta a}{\Delta N} = 6 \left(\frac{\Delta K}{E} - 2.7 \times 10^{-5} \right)^2 \frac{K_C}{K_C - \Delta K} \qquad (2)$$

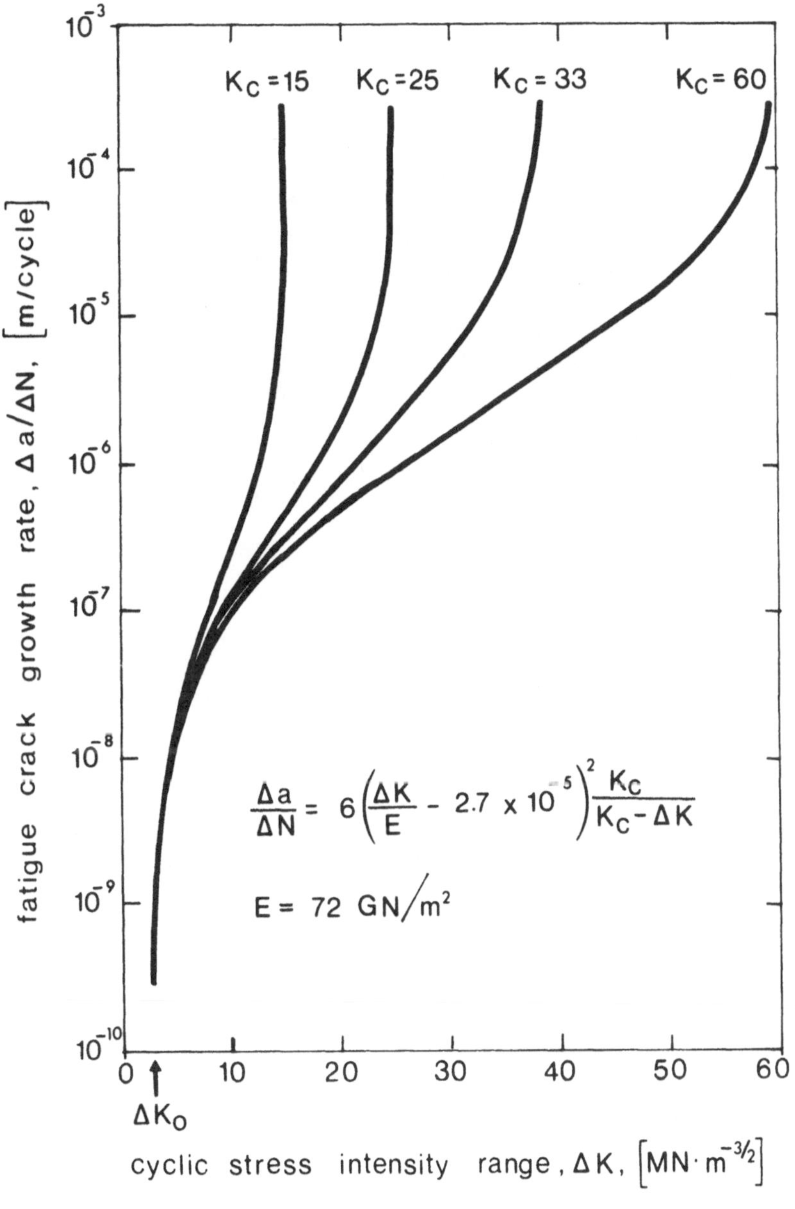

Fig. 9   Theoretical prediction of the growth rate of
         fatigue cracks in aluminum alloys of different
         fracture toughness

where a is the crack length, N the number of load cycles, and E is the modulus of elasticity. Equation (2) has been derived on the assumption that fatigue crack growth is due to alternating shear and that the growth rate per load cycle is determined by the crack tip opening displacement[29]. This leads to a dependence of the fatigue crack growth rate

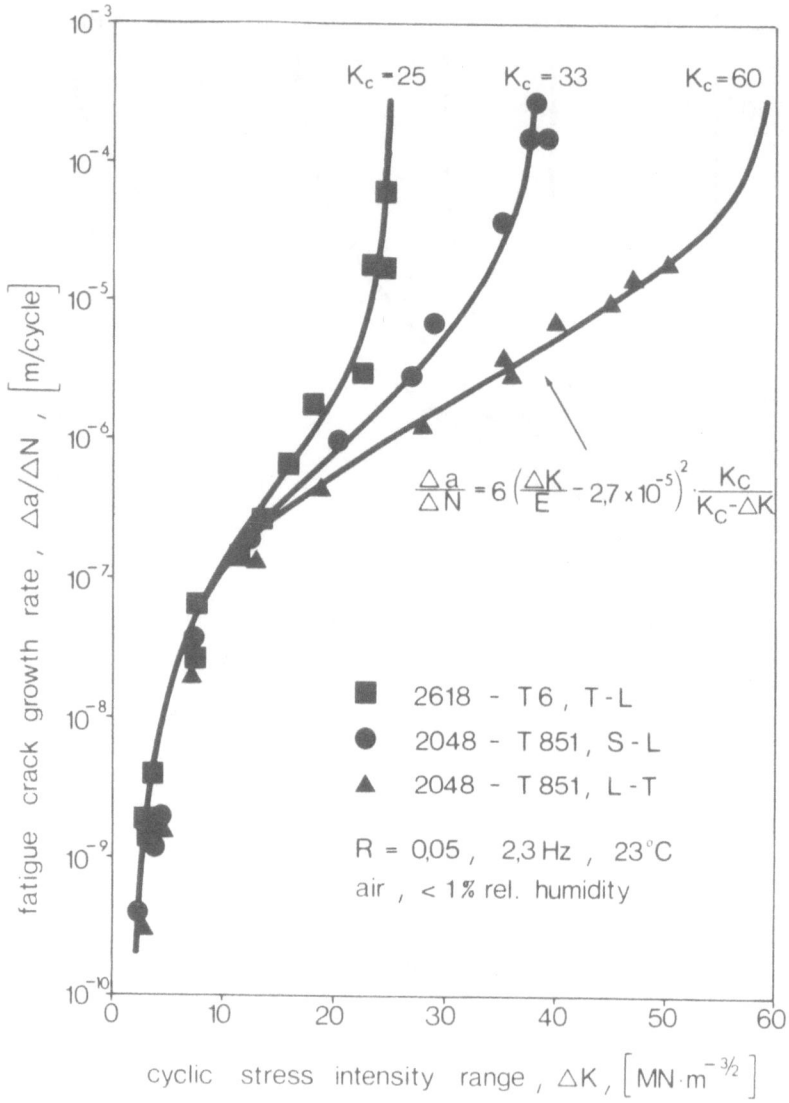

$$\frac{\Delta a}{\Delta N} = 6 \left(\frac{\Delta K}{E} - 2.7 \times 10^{-5} \right)^2 \cdot \frac{K_C}{K_C - \Delta K}$$

■   2618 - T6, T-L

●   2048 - T851, S-L

▲   2048 - T851, L-T

R = 0.05, 2.3 Hz, 23°C

air, < 1% rel. humidity

Fig. 10   Comparison of experimentally measured and theoretically predicted growth rates of fatigue cracks in aluminum alloys

on ΔK square. The second term in the brackets is an empirical correction for the existence of a fatique threshold stress intensity. In agreement with equation (2), ΔK_o can be determined to a first approximation by the following empirical relation[5)25)]:

$$\frac{\Delta K_o}{E} = (2.7 \pm 0.3) \cdot 10^{-5} \sqrt{m} \qquad (3)$$

The K_c correction in equation (2) is a simplified acknowledgement of the fact that $\Delta a/\Delta N$ raises steeply as ΔK approaches K_c or K_{Ic}. There are many other possible and perhaps more sophisticated equations to describe the fatigue crack growth rate as influenced by stress intensity and materials properties, but equation (2) is not only simple, but also reasonably successful in predicting the fatigue behavior of widely different materials[28], an example being provided by the data in Fig. 10.

Stress corrosion cracking

The resistance to stress corrosion crack growth is shown in Fig. 11 and Fig. 12 for the older and the newly developed alloys of the 2XXX (Al-Cu-Mg) and, the 7XXX (Al-Zn-Mg-Cu) series[3]. A detailed comparison reveals that it is not just the different alloy composition which renders the newer allovs more resistant to stress corrosion cracking. Rather, it is the heat treatment which is most important, specifically the overaging practice of the -T7, T736 and -T73 variety.

Among the 2XXX series alloys with copper as the major alloying element, the effect of aging practice is exemplified by alloy 2219, Fig. 11. In the -T37 temper, it belongs to the most stress corrosion susceptible alloys of all; in the artificially aged -T87 temper it is completely immune to SCC in the salt water test, i.e. its K_{Iscc} in the -T87 temper is as high as its fracture toughness. It can also be seen in Fig. 11 that the new alloy 2048-T851 (thick plate) has a K_{Iscc} value as high as the fracture toughness of some of the older alloys.

As Fig. 12 shows, neither the composition nor the heat treatment can render the Al-Zn-Mg-Cu alloys entirely immune to stress corrosion cracking. For example, alloy 7475-T651 exhibits a stress corrosion crack growth rate which is very similar to the one of alloy 7075-T651. However, the addition of at least 2% Cu , combined with a suitable amount of overaging

(-T736, -T73) reduces the growth rates of stress corrosion cracks by a very significant factor. This has resulted in a great decrease of the number of actual stress corrosion service failures with high strength aluminum alloys. As an added benefit, overaging also helps to prevent exfoliation corrosion.

The detailed atomistic mechanisms by which intergranular stress corrosion cracking in aluminum alloys can occur are still debated and not yet understood; there is not enough space here to present the arguments[1)-21)]. However, it appears that both, chemical and mechanical influences can be decisive. While certainly overaging will reduce the chemical differences between the grain boundary area and the grain interior, it is also of interest to note that overaging a high purity 7075 material considerably reduces the intergranular component of the mixed fracture mode[30)], thus indicating that it may be more difficult for intergranular cracks to propagate in overaged 7XXX alloys.

Figures 11 and 12 sum up a long and successful effort in America to develop high strength aluminum alloys of superior resistance to stress corrosion cracking. While the basic atomistic mechanisms at the crack tip are still subject to speculations and discussions, the alloys of the 2XXX series

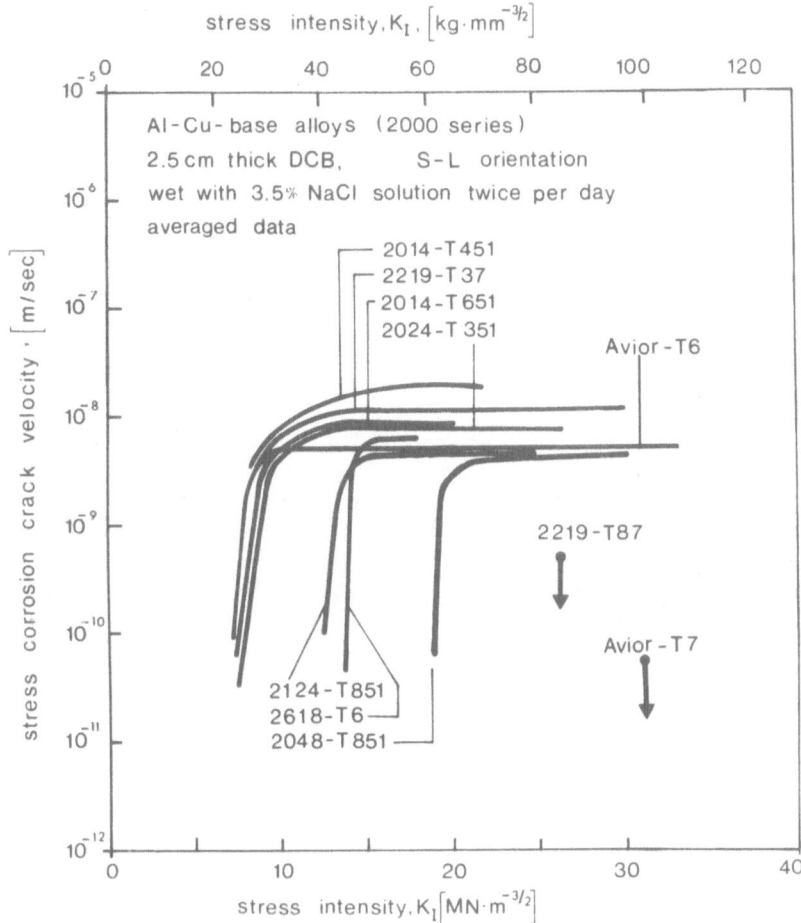

Fig. 11   Effect of stress intensity on the velocity of
          stress corrosion cracks in Al-Cu-Mg alloys of
          high strength.

(Al-Cu-Mg) and of the 7XXX series (Al-Zn-Mg-Cu) have been
dramatically improved over the last few years, simply by trial
and error, as well as by "enlightened empiricism". Fig. 13
summarizes this alloy development and shows that in the 2XXX
alloys K_{Iscc} was improved with little effect on the plateau
velocity of stress corrosion cracks. In the 7XXX series, it
was on the contrary, the plateau velocity of stress corrosion
cracks which was most significantly improved, and the effects

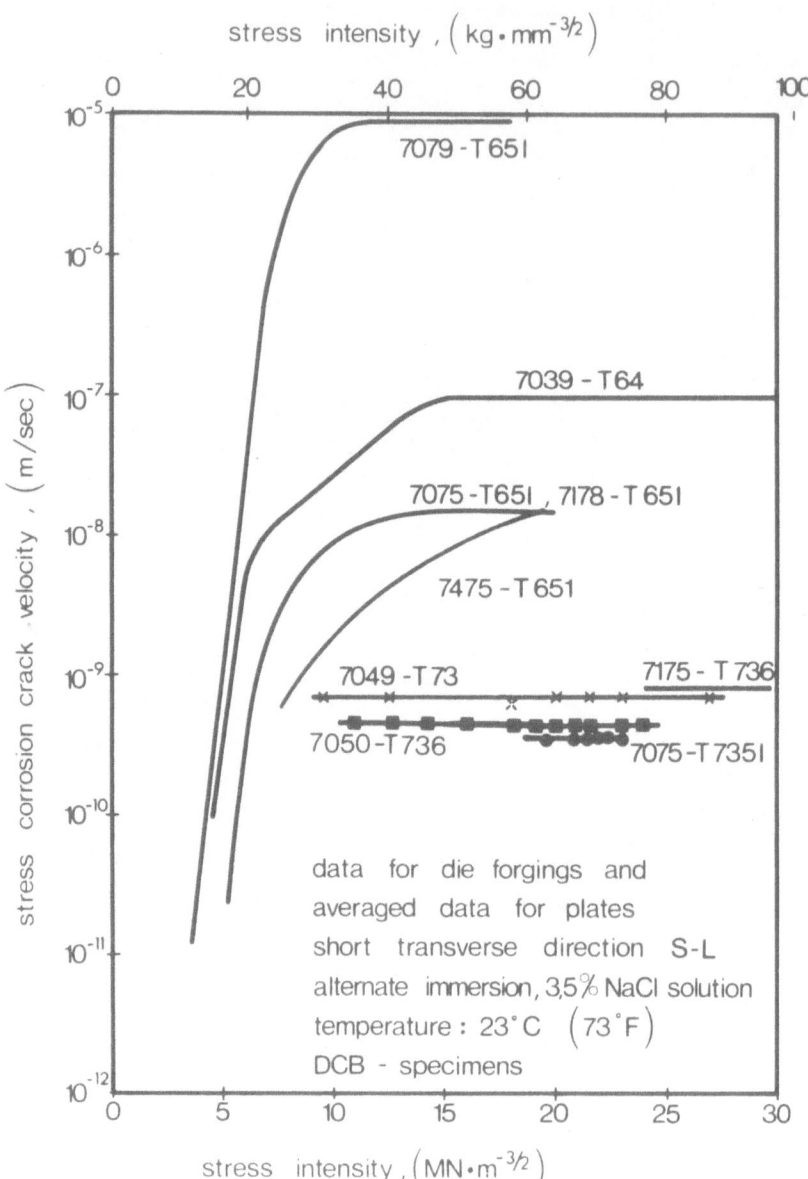

Fig. 12   Effect of stress intensity on the velocity of stress
           corrosion cracks in Al-Zn-Mg-Cu alloys of high
           strength

of alloy development on K_{Iscc} in the 7XXX series are unknown or debatable.

The improvement in stress corrosion resistance in commercial high strength aluminum alloys is not limited to tests in NaCl solution (Fig. 11 and Fig. 12). As can be seen in Fig. 14,

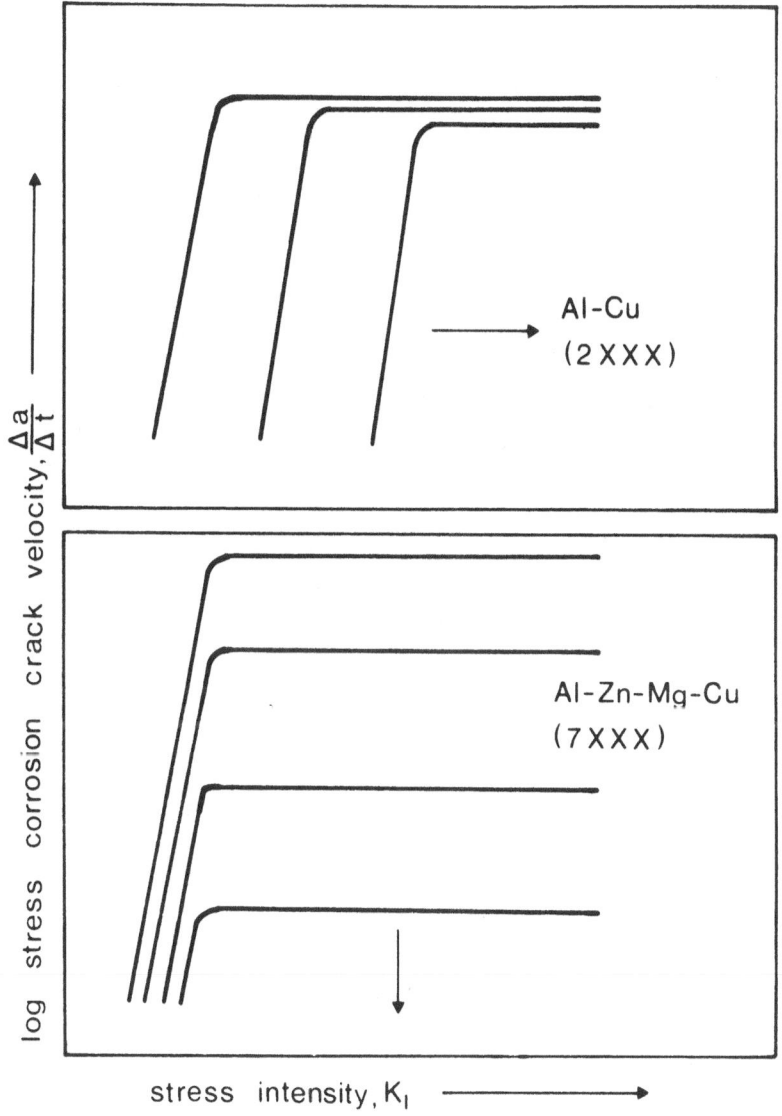

Fig. 13   Schematic representation of the improvement of
          the stress corrosion performance of two different
          groups of aluminum alloys

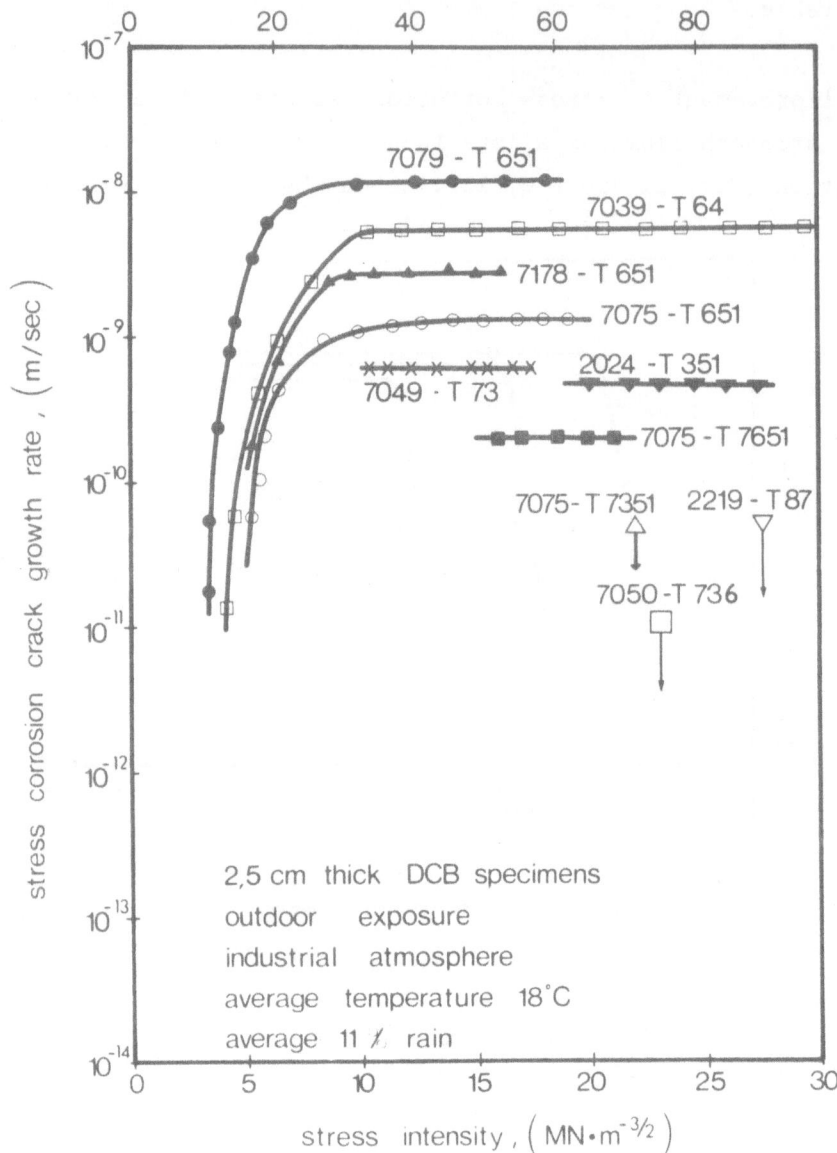

Fig. 14   Comparison of stress corrosion crack grwoth rates
           of high strength wrought aluminum alloys in an
           industrial atmosphere

the new and overaged aluminum alloys are also remarkably
more resistant to stress corrosion outdoors in industrial
atmosphere. Some of them are completely immune to SCC in this
environment.

The success in this American alloy development of high strength
stress corrosion resistant aluminum alloys is paralleled only
by the equally successful improvements in fracture toughness
of the same classes of alloys, as discussed above and shown
in Fig. 5 and Fig. 6.

Crack paths in stress corrosion and corrosion fatigue

In commercial high strength wrought aluminum alloys the
individual crystals are flattened and thus, the grain boundaries
are preferentially oriented in the plane which is determined
by the longitudinal and transverse directions, as illustrated
in Fig. 15. Since stress corrosion cracks in aluminum alloys
follow almost exclusively the grain boundaries, by far the
largest number of stress corrosion failures is due to stresses
in the short transverse direction, i.e. with crack orientations
S - L and S - T, where the tensile stress is perpendicular to
the grain boundaries. The typical intergranular fracture path
of SCC in aluminum alloys is shown in Fig. 16.

In contrast to stress corrosion cracks, fatigue cracks and corrosion
fatigue cracks are normally transgranular as shown in Fig. 17
where a typical corrosion fatigue crack is seen to cut across
several grains. The difference between the intergranular
stress corrosion crack in Fig. 16 and the corrosion fatigue
crack in Fig. 17 is characteristic for the two different
kinds of environment-assisted subcritical crack growth.

But there are exeptions, as shown in Fig. 18 and Fig. 19.
In Fig. 18, a corrosion fatigue crack is seen to follow a
suitably oriented grain boundary, thus giving an example of
intercrystalline corrosion fatigue. Fig. 19 shows that corrosion

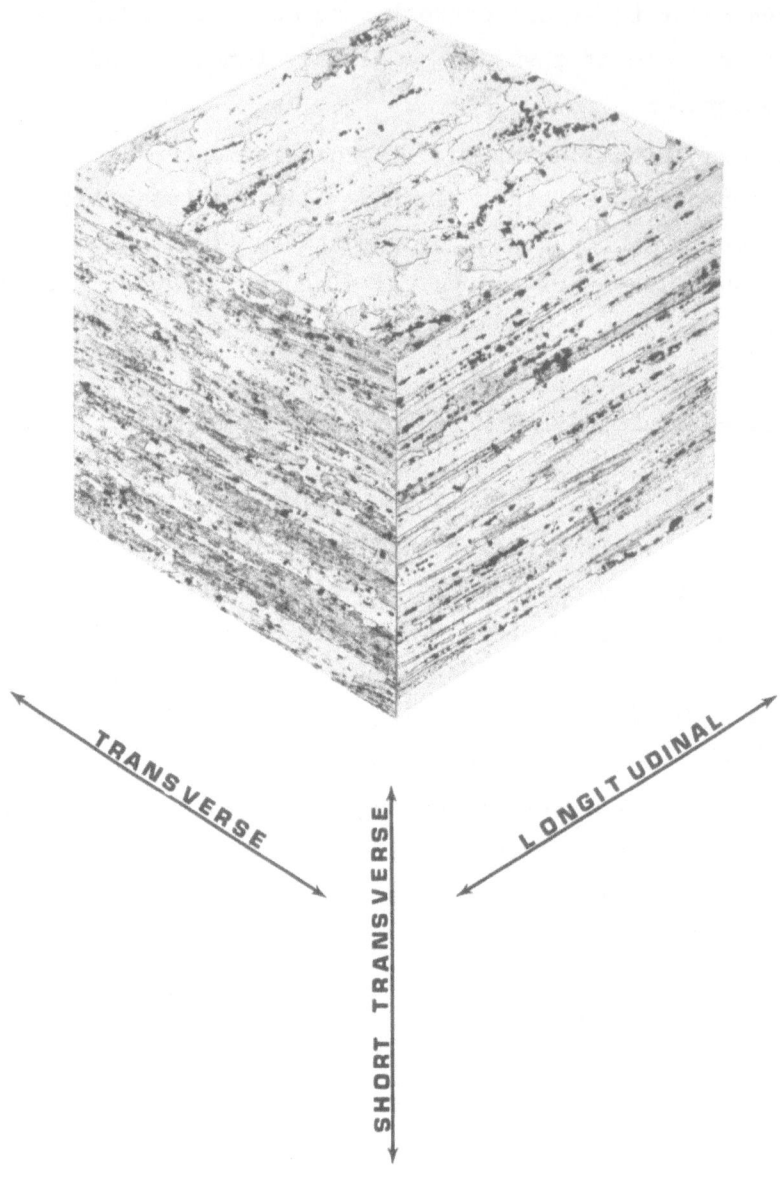

Fig. 15   Composite micrograph, showing the preferred
           orientation of grain boundaries with respect to
           the processing directions of plate material.
           Aluminum alloy 7079-T651

fatigue can be both <u>inter</u>granular and <u>trans</u>granular, and that the crack can exhibit many microscopic branches and an irregular crack front.

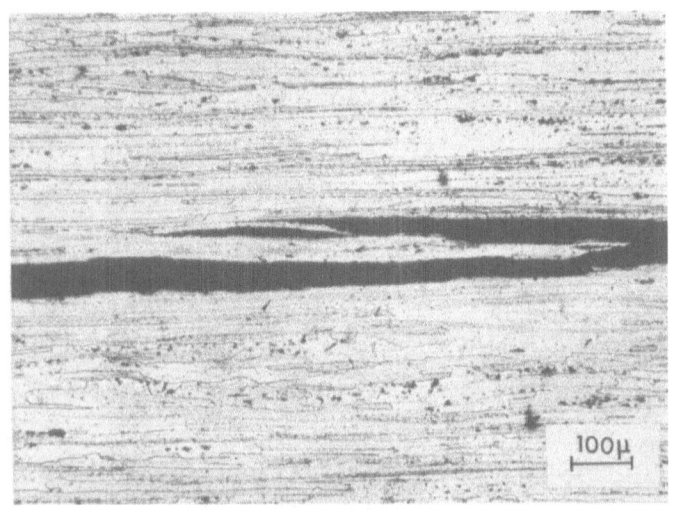

Fig. 16   Intergranular stress corrosion crack path in high strength aluminum alloy 7079-T651 (plate) exposed to NaCl solution

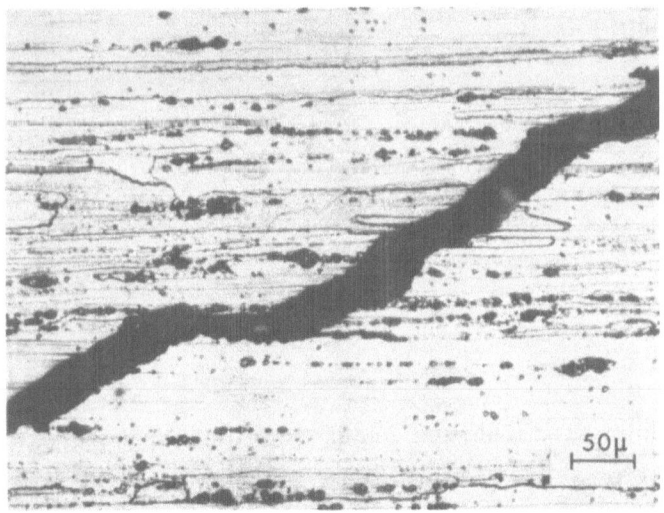

Fig. 17   <u>Trans</u>granular corrosion fatigue crack in high strength aluminum alloy 7079-T651

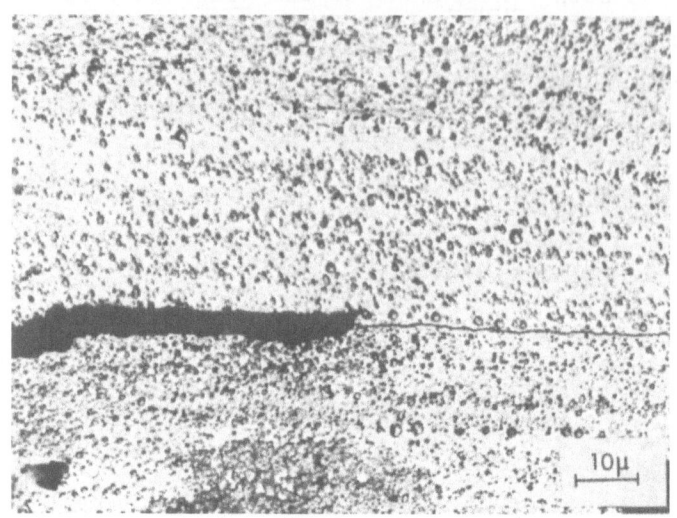

Fig. 18   Intergranular corrosion fatigue crack in high
strength aluminum alloy 7079-T651

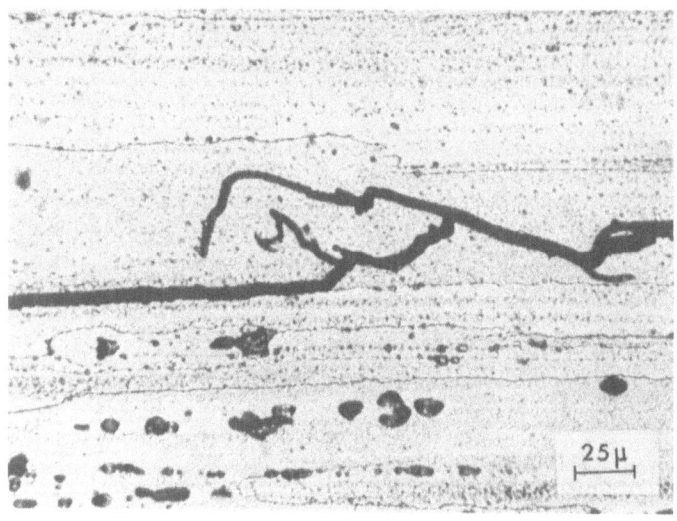

Fig. 19   Corrosion fatigue crack in high strength aluminum
alloy 7079-T651. Partly intergranular, partly
transgranular with irregular crack front and
microbranching

Based on the general rule that stress corrosion cracking in aluminum alloys is almost exclusively intergranular, and fatigue as well as corrosion fatigue is mostly transgranular, this characteristic difference can be useful in failure analysis. However, there are important exeptions, specifically if stress corrosion and corrosion fatigue occur together. This is discussed below.

"Stress corrosion cracking under cyclic loads" and "true corrosion fatigue"

The interaction between stress corrosion cracking, fatigue, and corrosion fatigue is a subject which has been found worthwhile of detailed investigations over the past years. This is because of the obvious technological importance as well as because of the many challenging questions it poses and also because the observed phenomenology might serve as guidance for the study of other interactions of time-dependent and cycle-dependent phenomena which are more difficult to study, such as creep and high temperature fatigue.

In the following, we will present a full set of data on stress corrosion, fatigue, and corrosion fatigue crack growth data for the aluminum alloy 7079-T651 with cracks in the S -L orientation. A comparison of such data with the predictions of some theoretical models and hypotheses leads to a refinement of our understanding of corrosion fatigue and stress corrosion under cyclic loads.

The stress corrosion crack velocity of alloy 7079-T651 is plotted as a function of stress intensity in Fig. 20. In saturated aqueous NaCl solution, the maximum crack velocity is typically around 10^{-5} m/sec. Now assume a fatigue test would be carried out with a precracked specimen of this alloy in saturated NaCl solution. Let us assume further that the load wave shape would be sinusoidal, zero to tension loading

148

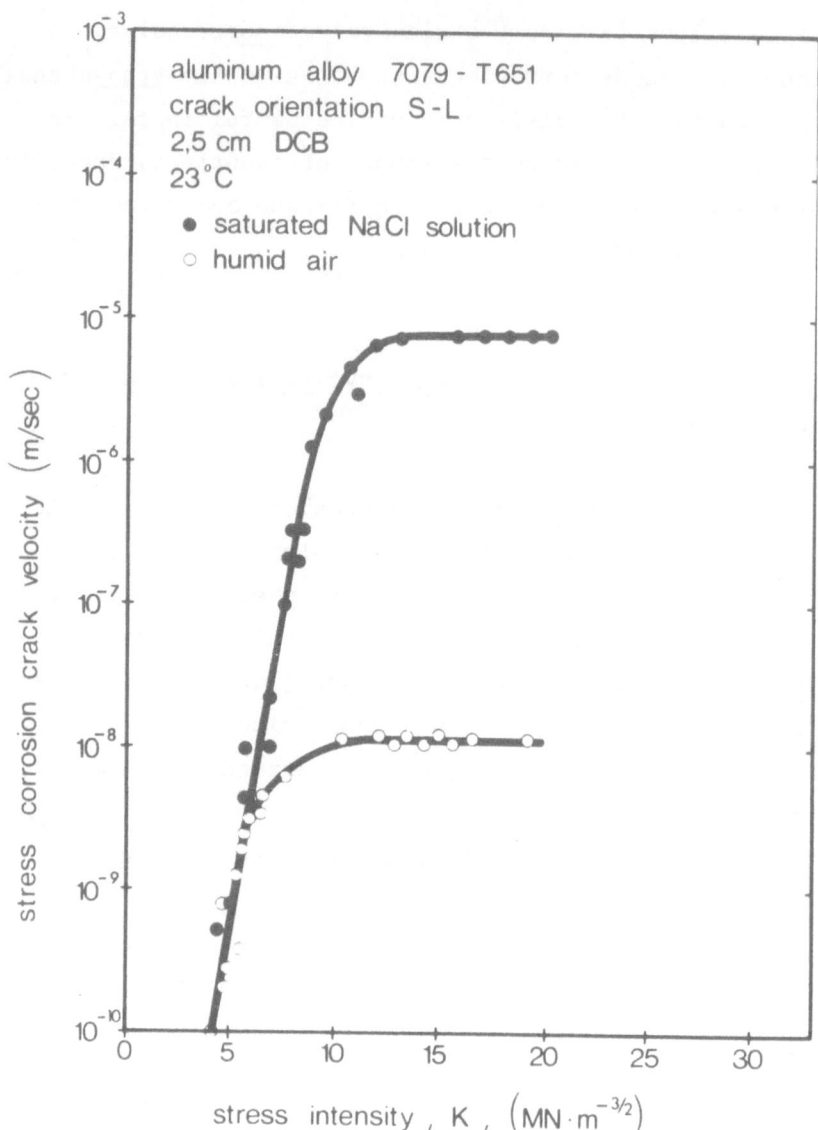

Fig. 20   Effect of stress intensity on the stress corrosion
crack velocity of a high strength aluminum alloy
in air and in aqueous NaCl solution

thus R ($\equiv K_{min}/K_{max}$) = 0, and with ΔK larger than K_{Iscc}.
In that case we would get a contribution of SCC to the overall
growth of the fatigue crack per load cycle. Wei and Landes[31]
have proposed that under such conditions the corrosion fatigue
crack growth rate in an aggressive environment can be considered
as the sum of two components; one representing crack extension
under cyclic loading in the absence of an aggressive environment, and the second representing crack extension due to
stress corrosion cracking.

Formally, the model is expressed by the equation[31][32]

$$(\Delta a/\Delta N)_c = (\Delta a/\Delta N)_r + \int (da/dt) \cdot k(t)\, dt \qquad (4)$$

where $(\Delta a/\Delta N)_c$ and $(\Delta a/\Delta N)_r$ represent the per cycle crack
extension due to cyclic loading in corrosive and inert
environments, respectively, and the integral represents the
sustained load crack extension during the time for a single
fatigue loading cycle. This integral is used to account for
the effects of frequency, mean load, range of cyclic loads, and
the load-time wave form on the sustained load growth component,
assuming da/dt is related to the instantaneous value of the
stress intensity factor during cyclic loading, K(t).

It was thought that this approach would allow predictions of
corrosion fatigue crack growth rates simply by adding the
(inert) fatigue crack growth per cycle and the crack extension
due to stress corrosion. This would obviously reduce the work
presently necessary to measure corrosion fatigue crack growth
rates. Partly this model is indeed useful, especially under
those conditions where the stress corrosion component is
significant, as in alloy 7079-T651. For example, the reduction
of the cyclic load frequency by a factor of ten shold increase

the stress corrosion component of the overall crack extension
per cycle by a factor of ten. Thus, if the stress corrosion
component dominates, $\Delta a / \Delta N$ in a corrosion fatigue test shold
depend linearly on frequency. This is indeed what we observe
for alloy 7079-T651 at high ΔK and low frequencies, as shown
in figure 21.

For the simple load wave shape mentioned above, we can
graphically integrate the SCC function $da/dt = f(K)$ which
is presented in Fig. 20 for 7079-T651 in NaCl solution.
According to equation (4) this should give us a prediction
of the corrosion fatigue crack growth rate which we could
compare to actually measured data. The predicted crack growth
rates are shown in Fig. 22 and the comparison with actual
data is made in Fig. 23 where the curves drawn are identical to
those in Fig. 22. In order to make the prediction of corrosion
fatigue crack growth data in Fig. 22, we need not only the
stress corrosion data presented in Fig. 20, but also, according
to equation (4), we need some information on fatigue crack
growth in inert environments. For this we have used equation (2)
as indicated in Fig. 22. However, as soon as stress corrosion
crack growth becomes dominant, especially at low frequencies
and high ΔK's, the inert-environment fatigue contribution
becomes insignificant and can be omitted. Thus, the predictions
made in Fig. 22 for corrosion fatigue represent just the
integral in equation (4).

The comparison between "theory" and experiment in Fig. 22
reveals one surprising and two expected results:

> The fatigue crack growth in vacuum is reasonably
> well predicted by equation (2). This is expected
> because of the good predictive capability of equation (2)
> which was already evident in Fig. 9 and Fig. 10 as
> well as with quite different materials[28].

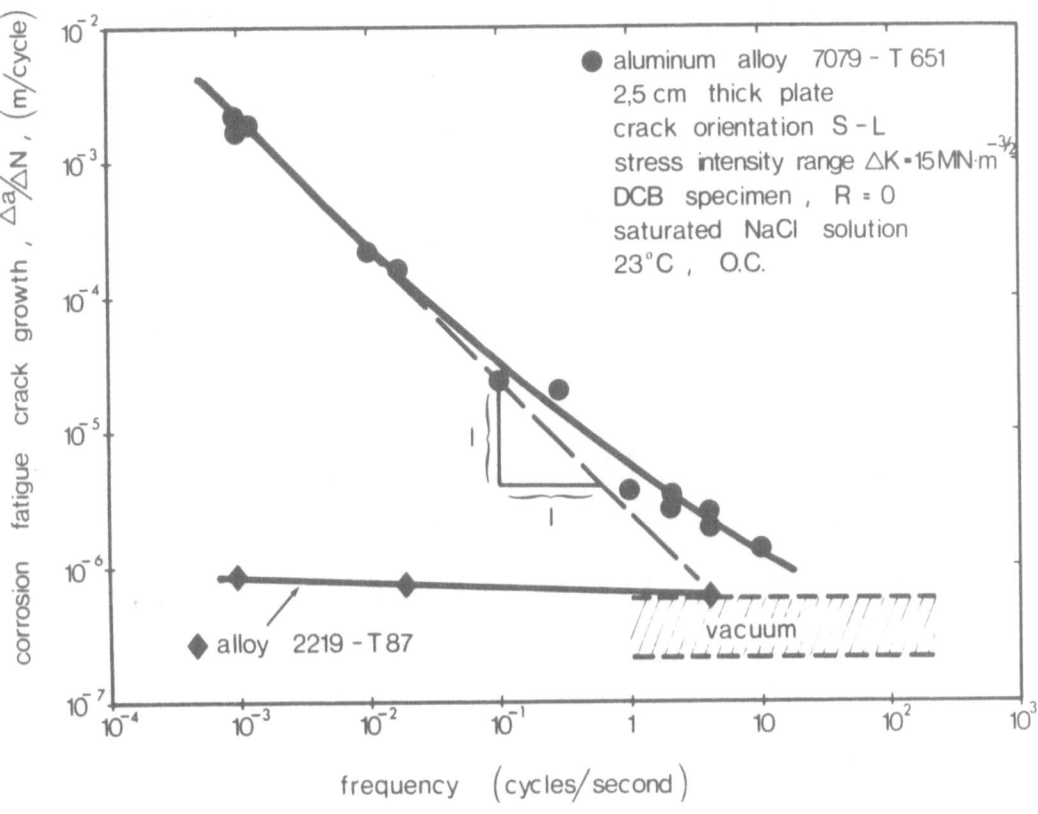

Fig. 21   Effect of frequency on the corrosion fatigue crack
           growth rate of two different high strength aluminum
           alloys

The corrosion fatigue crack growth rate at frequencies
from 10^{-1} Hz to 10^{-3} Hz is reasonably well predicted
from the knowledge of the stress corrosion crack
growth rate data (Fig. 20) and their integration
based on the known load wave shape. This, too, was
expected, because of the success of the Wei-Landes
superposition model with other materials[33] and
because the proportionality between frequency and
corrosion fatigue crack growth rate which is evident
in Fig. 21. The slope of -1 in that figure shows

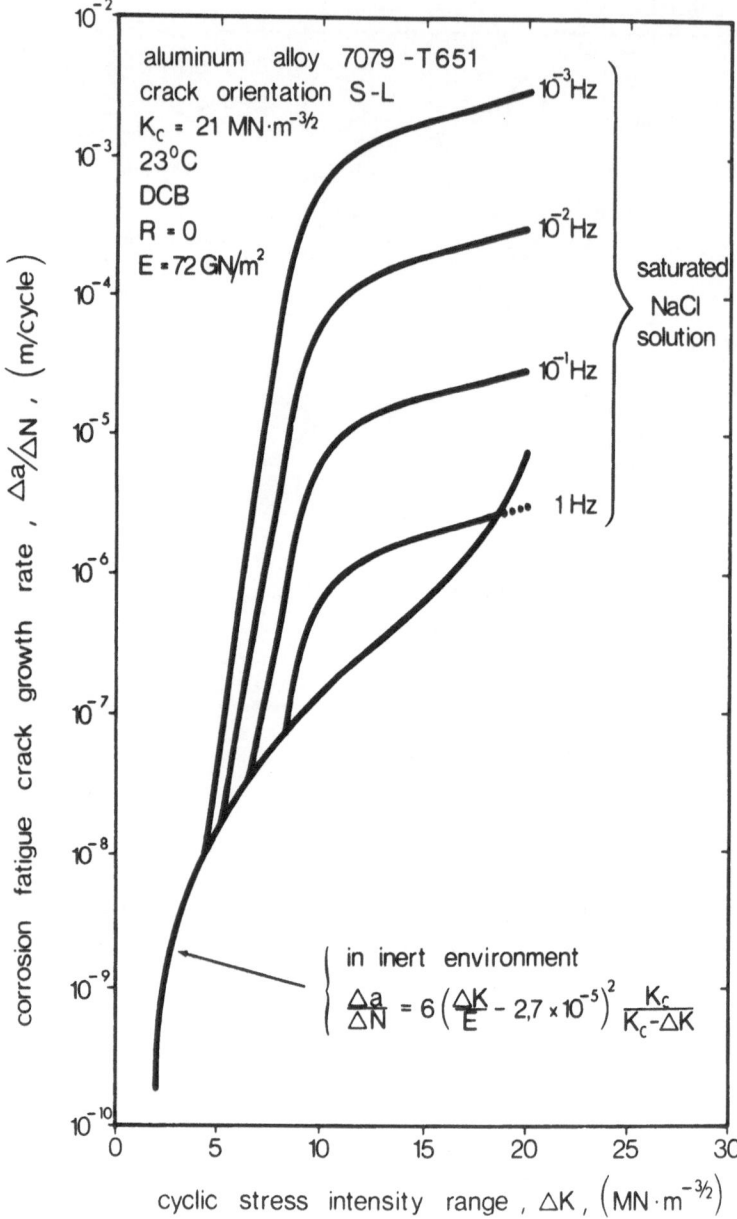

Fig. 22   Predicted fatigue and corrosion fatigue crack
           growth rates, based on equations (2) and (4)
           and on the SCC data presented in Fig. 20

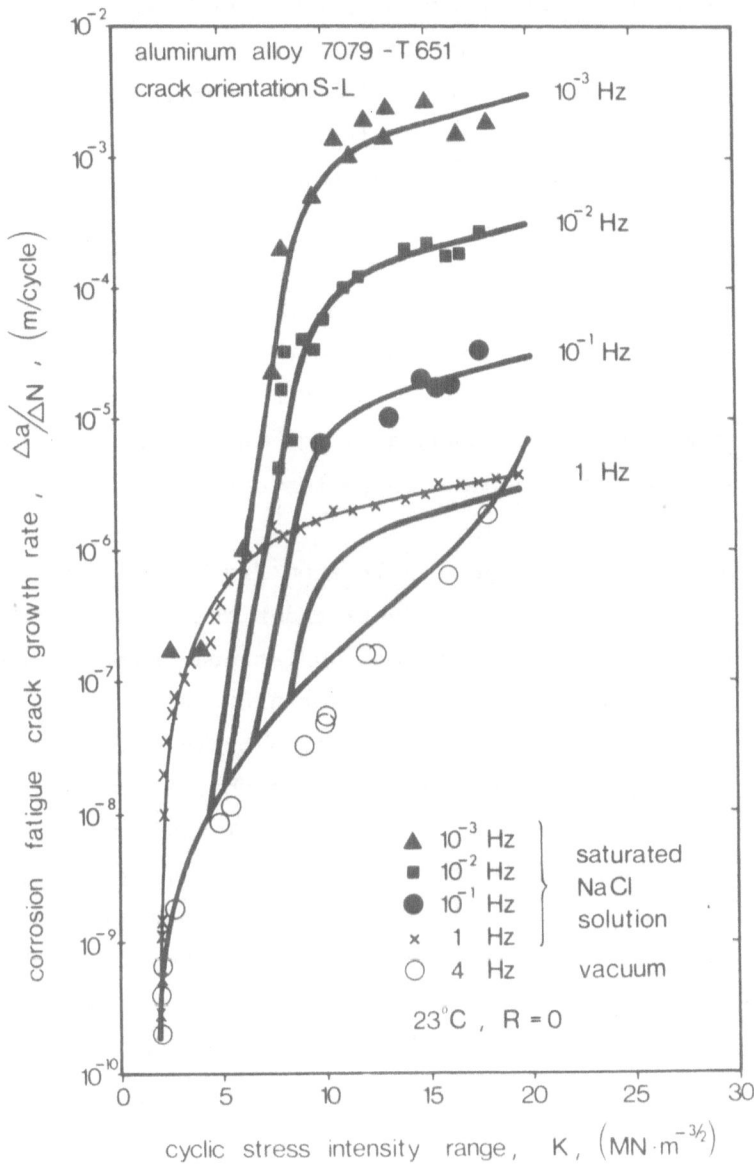

Fig. 23   Comparison between theory and experiment. The fat
          black lines are theoretical predictions as
          presented in Fig. 22

that corrosion fatigue crack growth in alloy 7079-
T651 is a time-dependent phenomenon, rather than a
cycle-dependent phenomenon. Thus the corrosion
fatigue crack growth at low frequencies and high cyclic
stress intensity ranges in this alloy is really a
kind of <u>stress corrosion under cyclic loads</u>, and
this is confirmed by the good fit of prediction
(based on SCC) and experiment in Fig. 23. This
observation of stress corrosion under cyclic loads
is supported by metallographic evidence: the corrosion
fatigue cracks at very low frequencies are almost
completely intergranular (Fig. 18) as one would
expect from a stress corrosion crack.

The one surprising result from Fig. 23 is that at
frequencies around 1 Hertz and higher a much stronger
effect of the environment on fatigue crack growth
is observed than one would predict from a summation
of fatigue and stress corrosion crack extension
during each load cycle. This is what we call "true"
corrosion fatigue. In alloy 7079-T651, stressed in
the short transverse direction, (S-L), "true" corrosion
fatigue is particularly strong, and can enhance the
crack growth rates up to a hundred times over those
observed in vacuum at intermediate ΔK. In this
particular alloy, "true" corrosion fatigue is partly
transgranular and partly intergranular (Fig. 19) and
the great acceleration of cracks is unusual. The
reason for this is that we have tested this highly
environment sensitive alloy in an orientation (S-L)
where extremely flat grain boundaries are preferentially
oriented in the plane of the crack. In other aluminum
alloys and particularly in other orientations the
fracture path due to "true" corrosion fatigue is
entirely transgranular and the effect on crack growth
rate is less, as will be discussed below.

Figures 24 and 25 constitute an attempt to sort out
the various contributions to corrosion fatigue crack
growth evident in Fig. 22, and to define the differences
between "fatigue crack growth", "true corrosion
fatigue", and "stress corrosion under cyclic loads".

Fatigue. In a narrow sense of the word, "fatigue"
is crack nucleation and growth under cyclic loads
in inert environments, such as vacuum or dry argon,
because other environments can assist cracking and
would then cause corrosion-fatigue. Fatigue in inert
environments is always indicated by the lowest curve
in the diagrams of Fig. 22, Fig. 23, Fig. 24, and
Fig. 25, corresponding to the lowest crack growth
rate at a given ΔK. Fatigue in inert environments is
reasonably well predicted by equation (2) for a
number of different materials, including aluminum
base alloys. It is thought that fatigue crack growth
is due to alternating shear at the crack tip, and
that the fatigue crack growth rate is controlled by
the crack tip opening displacement. When the cyclic
stress intensity range approaches K_{Ic}, crack extension
in fatigue is not limited to alternating shear;
other mechanisms of fracture like void growth and
coalescence, rupturing of enclusions, cleavage, and
grain boundary separation can contribute to the
overall crack growth per load cycle. This leads to
the upswing of the $\Delta a/\Delta N$ curves near K_{Ic} or K_c
which is approximated in equation (2) by the K_c correction.
Fatigue in inert environments is characteristically
transgranular at intermediate and low ΔK values.

Stress corrosion under cyclic loads is a new term
which we would like to establish because we think
it serves the useful purpose of describing a phenomenon
which is different from other forms of corrosion

156

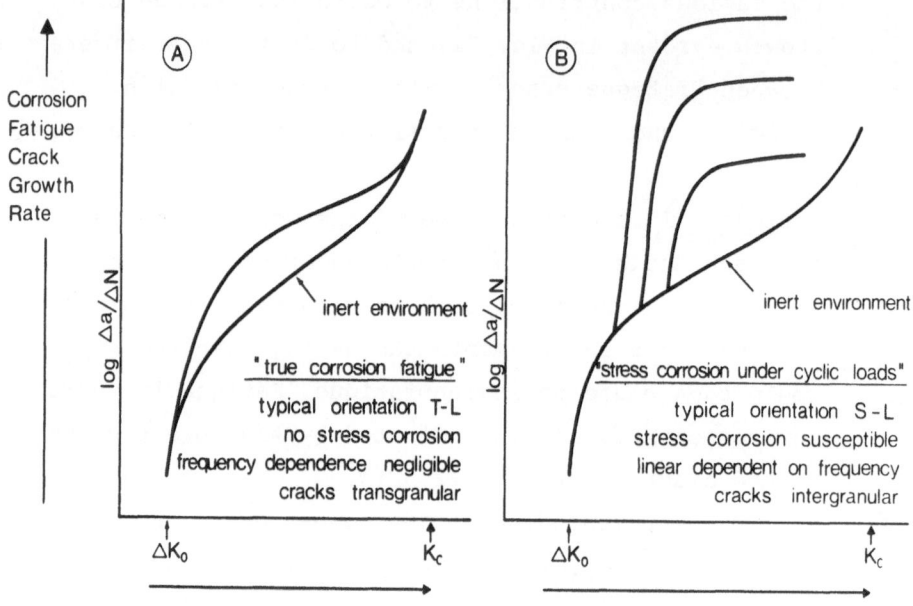

Fig. 24   Two types of environment-assisted fatigue crack
          growth: "true corrosion fatigue" and "stress
          corrosion under cyclic loads"

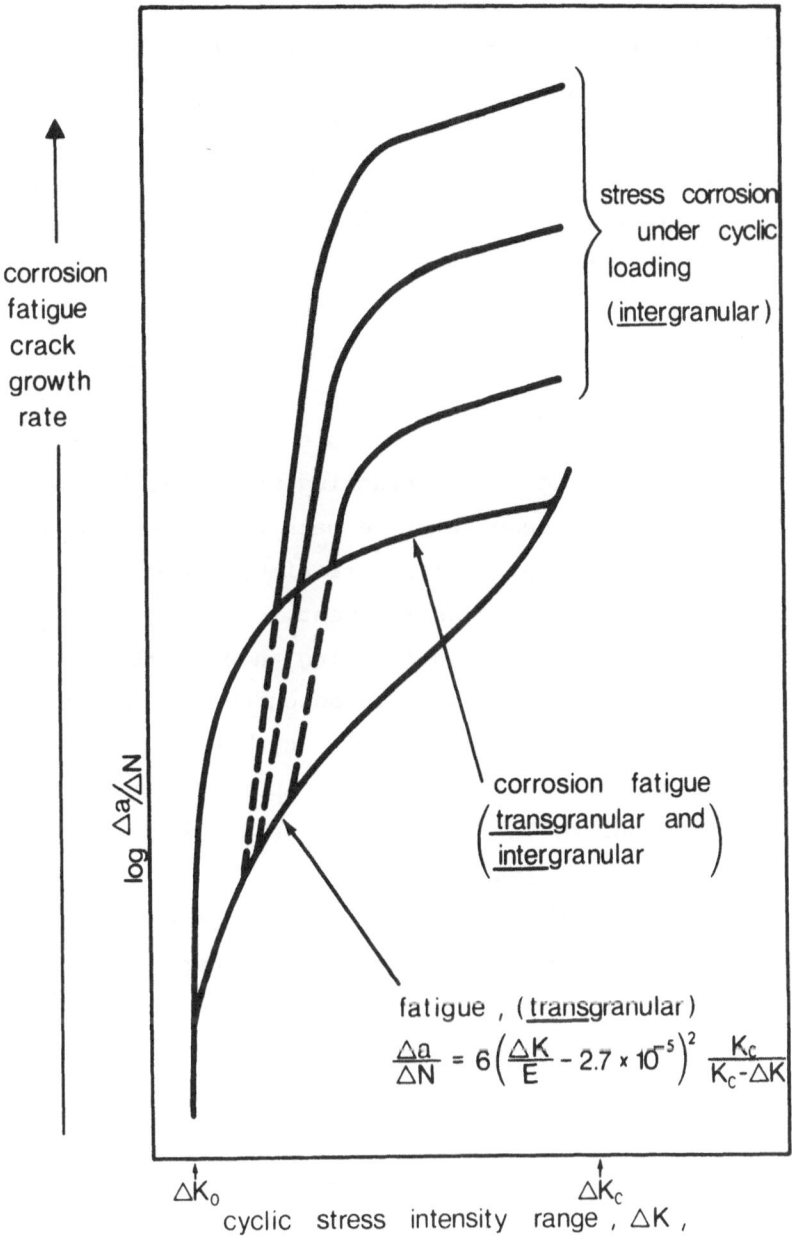

Fig. 25   Interpretation of the predicted and observed
           crack growth rates from Fig. 22 and Fig. 23

fatigue. Stress corrosion under cyclic loads refers to the kind of crack growth which is dominated by stress corrosion crack extension during the tensile part of a load cycle. By stress corrosion, we mean generally all types of "sustaines load crack growth"; in aluminum alloys (and in most high strength steels [28][33][34]) this is intergranular (Fig. 16, Fig. 18). In other alloy-environment combinations stress corrosion under cyclic loads could conceivably be transgranular. A number of characteristics of stress corrosion under cyclic loads are as follows:

1) Stress corrosion cracking under cyclic loads has always the same fracture path as stress corrosion cracking, or sustained load cracking, of the same material in the same environment.

2) Stress corrosion cracking under cyclic loads is time-dependent, not cycle-dependent. Thus, where stress corrosion cracking under cyclic loads pre-dominates, the crack growth rate per load cycle is proportional to the inverse of the loading frequency, as shown by the line of slope -1 in Fig. 21. This is also shown by the observed crack growth rate data in Fig. 23, which are spaced by a factor of 10 from each other for 10^{-1} Hz, 10^{-2} Hz, and 10^{-3} Hz, respectively.

3) Stress corrosion cracking under cyclic loads can be quantitatively predicted for relatively high ΔK's and low frequencies, as shown by a comparison of Fig. 22 and Fig. 23. This is also true for iron base alloys[33][34].

4) Stress corrosion cracking under cyclic loads is not simply "corrosion fatigue above K_{Iscc}", since it can also extend to lower stress intensity ranges in specific cases[34], because "fatigue is a more effective process for creating new crack surfaces than sustained loading"[33]. In our specific case of aluminum alloy 7079-T651, however, stress corrosion

under cyclic loading occured only above K_{Iscc} (Fig. 20, Fig. 22).

True corrosion fatigue is the acceleration of fatigue cracks due to an environment under conditions where sustained load cracking (or stress corrosion) does not occur. True corrosion fatigue differs greatly from stress corrosion under cyclic load on at least two counts: it is by far less frequency dependent and it is predominantly transgranular. The acceleration of fatigue cracks by "true corrosion fatigue" is almost always observed with aluminum base alloys in aqueous solutions, even under conditions where no stress corrosion cracking occurs at all or where SCC would be by far too slow to account for the crack acceleration. Thus, true corrosion fatigue occurs in all crack orientations irrespective of the preferred orientation of the grain boundaries, owing to its transgranular nature (Fig. 17), although occasionally it has intergranular components (Fig. 19). The insensitivity of true corrosion fatigue to structure and crack orientation contrasts sharply with stress corrosion under cyclic loads which occurs only along grain bondaries and only under cyclic tensile stresses in the short transverse direction of high strength wrought aluminum alloys.

The very small frequency dependence of true corrosion fatigue is seen in Fig. 21 (alloy 2219-T87) and in Fig. 26, (alloy 7075-T73). This contrasts sharply with stress corrosion under cyclic loads where the crack growth rate is proportional to the inverse of the loading frequency (Fig. 21).

As seen in Fig. 23 and Fig. 25, true corrosion fatigue, because it hardly depends on frequency, can be faster than the (predicted) stress corrosion under cyclic loads (which depends strongly on frequency and therefore is slow at high

frequencies. Thus as seen in Fig. 23, at a frequency of
1 Hertz, "true corrosion fatigue" outruns "stress corrosion
under cyclic loads".

This shows that true corrosion fatigue can be dominating
up to ΔK values far in excess of K_{Iscc}. Thus the term

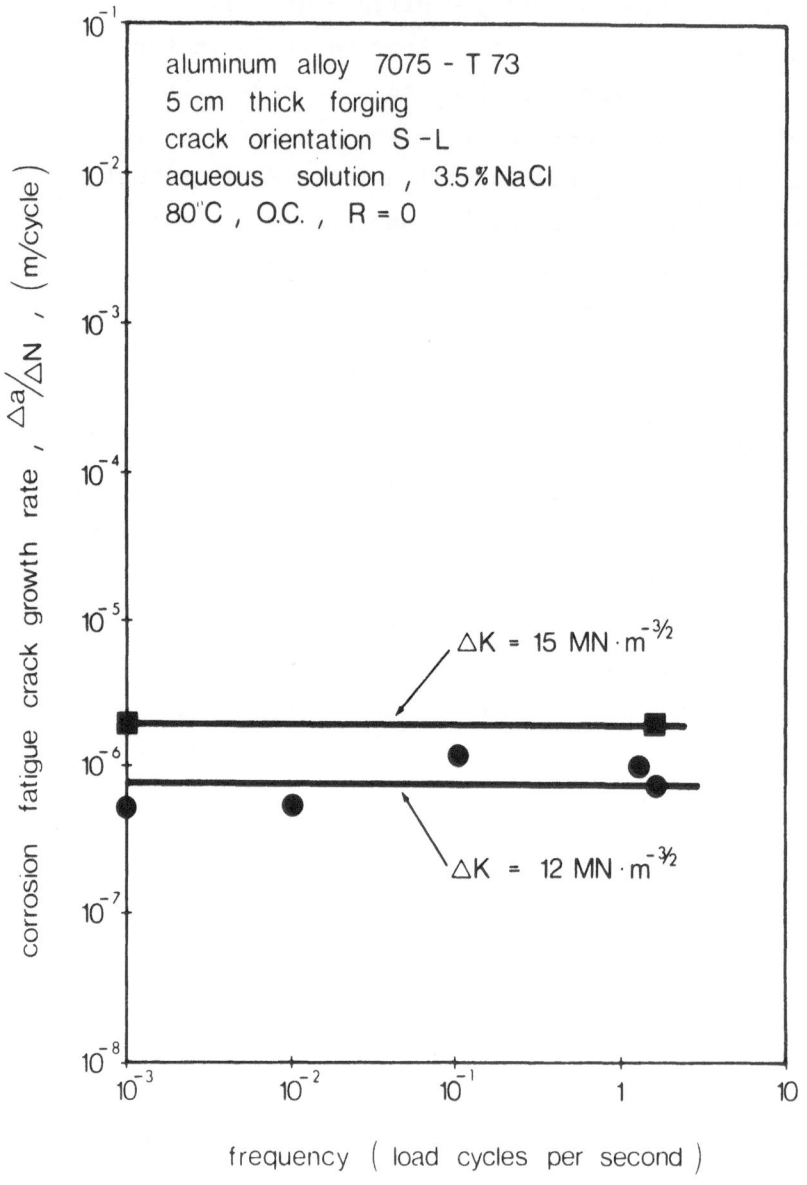

Fig. 26   Frequency-insensitive corrosion fatigue crack
           growth where stress corrosion is absent

"corrosion fatigue below K_{Iscc}" is not meaningful here: if the load frequency is high enough, the same environment-assisted, transgranular, frequency-insensitive, fatigue mechanism is observed above and below K_{Iscc}. This is why we prefer the term "true corrosion fatigue".

Fatigue crack growth in salt water

The growth rates of fatigue cracks in high strength aluminum alloys exposed to aqueous NaCl solutions are shown in Fig. 27 as functions of the cyclic stress intensity range, ΔK. Like in Fig. 7 the crack orientations were all L-T, and the load wave shapes were sinusoidal. The data presented in Fig. 27 are the centerlines of experimental scatterbands or "best fits" through the data in several published and unpublished recent reports.[27] [35]-[40]

Again, the experimental curves shown in Fig. 27 can be represented by a scatterband for the growth rate of fatigue cracks in aluminum alloys exposed to salt water. This scatterband is shown in Fig. 28, together with the summary scatterband for crack growth in air from Fig. 8. Fig. 28 allows a determination of the general effect that salt water has on the fatigue crack growth of high strength aluminum alloys cyclically loaded in the longitudinal direction.

All the corrosion fatigue cracking reported in Fig. 27 and Fig. 28 is transgranular and thus represents "true corrosion fatigue". It is seen that the effect of salt water on the growth rate of fatigue cracks is most marked at intermediate ΔK values. In the stress intensity range between 10 and 25 $MN \cdot m^{-3/2}$, the presence of salt water causes an average three to four times faster growth of fatigue cracks compared to the fatigue crack growth rate in aluminum alloys exposed to air. Some alloys, however, are much less affected by the presence of salt water[27]. For example, fatigue crack

growth rate data for alloy 2219-T87 in the L-T orientation
are almost indentical in both, air and salt water. In Fig. 28,
such data are represented by the narrow region of overlap
between the scatterbands for tests in air and in salt water.

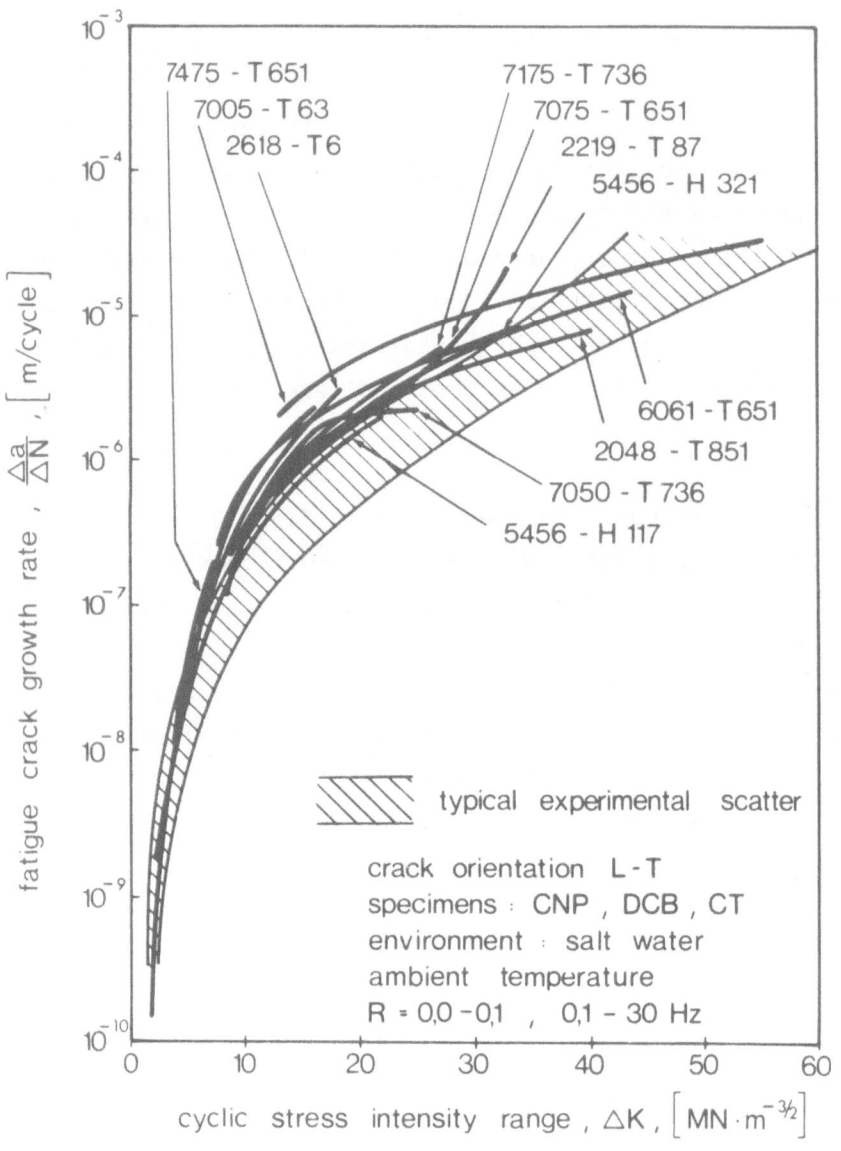

Fig. 27   Average growth rates of fatigue cracks in aluminum
           alloys immersed in 3.5% NaCl solutions

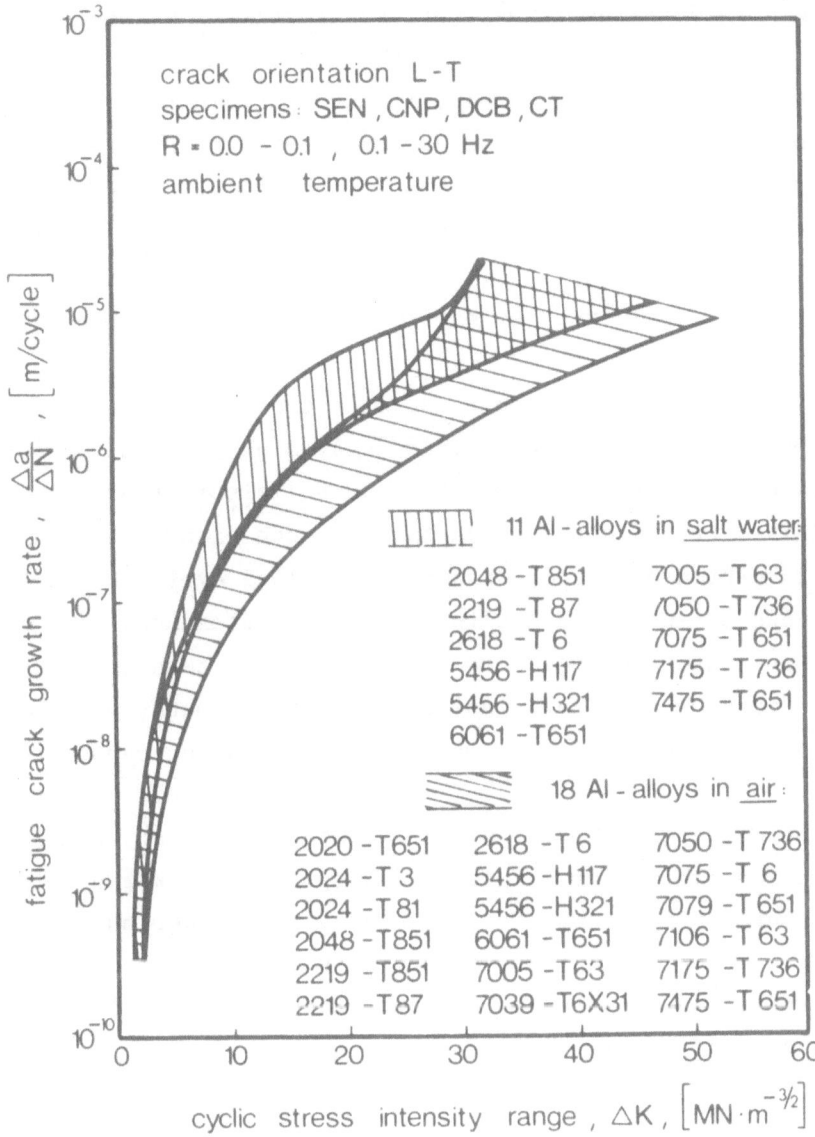

Fig. 28　Comparison of growth rates of fatigue cracks in aluminum alloys exposed to air and to salt water

At very low ΔK values, the effect of salt water on the fatigue crack growth rate vanishes. This confirms earlier observations that the stress intensity threshold for fatigue crack growth, ΔK_o, in aluminum alloys is not measurably

affected by the environment[5]. This phenomenon clearly
needs more detailed investigation, firstly to establish
more firmly the effect of environments on ΔK_o, and secondly
to evaluate the relationship, if any, between the stress
intensity threshold for fatigue crack growth, ΔK_o, and the
fatigue endurance limit of smooth specimens. Such investigations
appear all the more important, since the endurance limit
(or fatigue strength at 10^7 of 10^8 cycles) is well known to
be strongly influenced by enviroments such as salt water.

Fig. 28 also indicates that the effect of salt water on the
fatigue crack growth rate vanishes at high ΔK levels, where
ΔK approaches the fracture toughness. This is readily
explained by the observation that at such high ΔK levels a
major portion of the crack growth per load cycle is due to
void nucleation, void growth, and void coalesence. This mode
of crack extension does not depend on the external environment,
since most of it occurs inside the metallic material just
ahead of the crack front, where the environment cannot reach.

Recent alloy development has had almost no effect on true
corrosion fatigue crack growth, except for a lowering of
crack growth rates at high ΔK levels. This lowering of the
fatigue crack growth rate is due to improved fracture
toughness, and applies to fatigue crack growth in corrosive
environments as well as in air.

The effect of electrical potential on corrosion fatigue

The effect of potential on the corrosion fatigue crack growth
rate has been investigated for only a few aluminum base alloys
such as 7075-T651 and 7079-T651[5) 41) 42]. Possibility and
effectiveness of cathodic protection, can be estimated from
the data presented in Fig. 29, Fig. 30, and Fig. 31.

Fig. 29 shows corrosion fatigue crack growth rate results

which were measured under open circuit and the conditions
of "true corrosion fatigue". Obviously, chloride, bromide
and iodide solutions cause much faster crack growth rates
than distilled water, especially at intermediate ΔK values. The
upper two curves in Fig. 30 sum up the open circuit data
in KI and KBr solutions and the crack growth data in
distilled water. Superimposed on these curves are the crack
growth rate data measured in aqueous halide solutions
under cathodic protection. It appears that cathodic protection
can reduce the crack growth rates only down to those observed
in distilled water. This result, however might be particular
to the alloy investigated, since others report that in alloy
7075-T651 cathodic protection will reduce the corrosion
fatigue crack growth rate down to values normally observed
in dry argon[42].

The effect of potential on the corrosion fatigue crack growth
rate in the aluminum alloys 7079-T651 and 7075-T651 is
illustrated in Fig. 31 for constant cyclic stress intensity
amplitudes[5][41]. In alloy 7079, an intermediate crack growth
rate is observed under open circuit conditions. Anodic
polarization accelerates the cracks, cathodic polarization
retards it. The data for alloy 7075[41] appear to support
the statement[42] that cathodic protection can reduce the
crack growth rate to the level observed in dry argon. A
change in fracture plane orientation occurs when the potential
is changed significantly. Anodic potentials enhance the
crystallographic dependence of fatigue striation formation
by cleavage while cathodic protection results in ductile
striation formation by a shear process[42].

The just mentioned effects of potential refer to "true corrosion
fatigue crack growth". Concerning "stress corrosion under
cyclic loads" we can deduce from the well known effect of
potential on SCC[1] that cathodic protection would be
extremely effective in reducing the large growth rates

166

per load cycle observed at low frequencies in Fig. 23.
Since we know that cathodic protection can reduce the
SCC growth rates by a factor of one thousand in alloy
7079[1), the whole phenomenon of "stress corrosion under

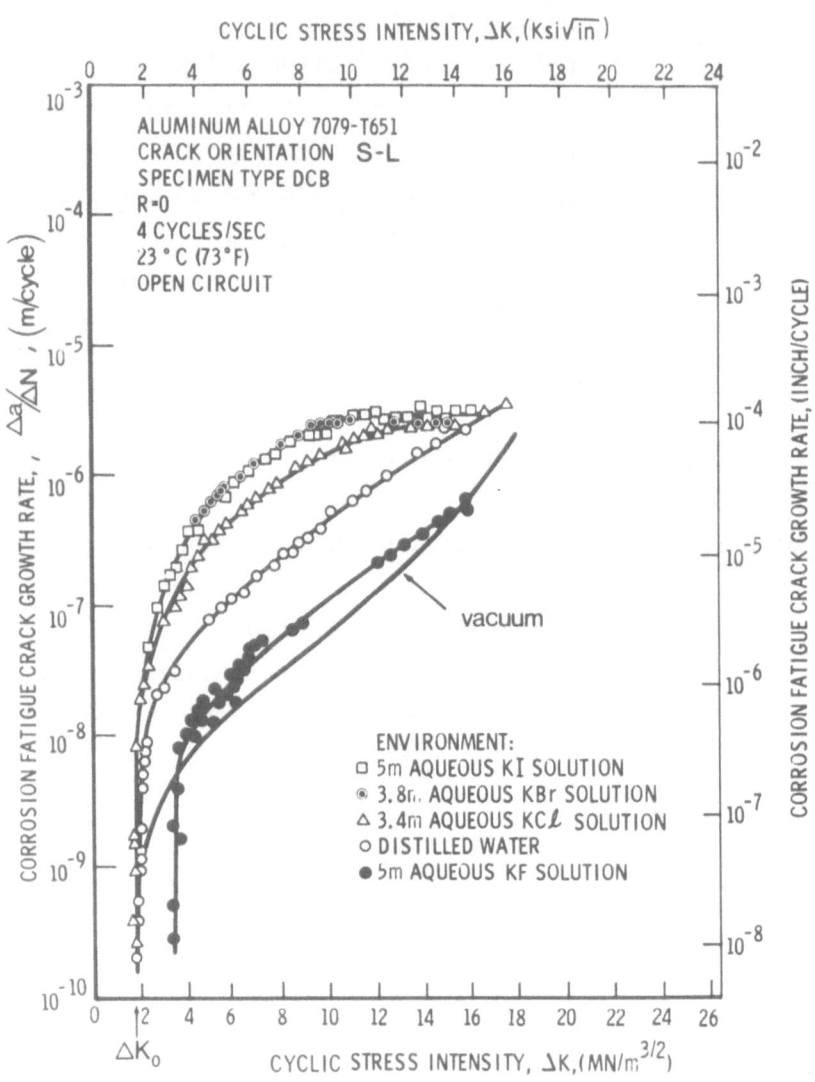

Fig. 29   Effect of cyclic stress intensity range on the
growth rate of corrosion fatigue cracks in a
high strength aluminum alloy exposed to various
environments

cyclic loads", which is so evident in Fig. 21, Fig. 22, Fig. 23, and Fig. 25, would simply disappear if the alloy were cathodically protected.

In practice there are of course limits to cathodic protection of aluminum alloys, one of the limits being given by the danger of "overprotection" which could lead to caustic attack[4].

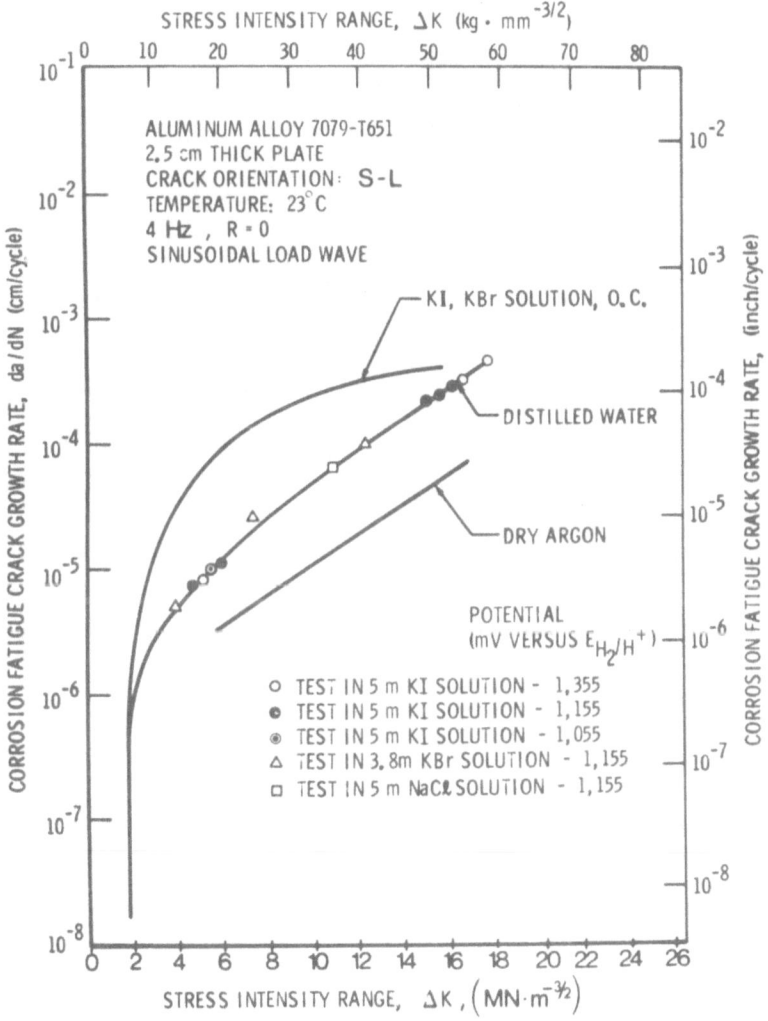

Fig. 30   Effect of cathodic protection on the growth rate of corrosion fatigue cracks in alloy 7079

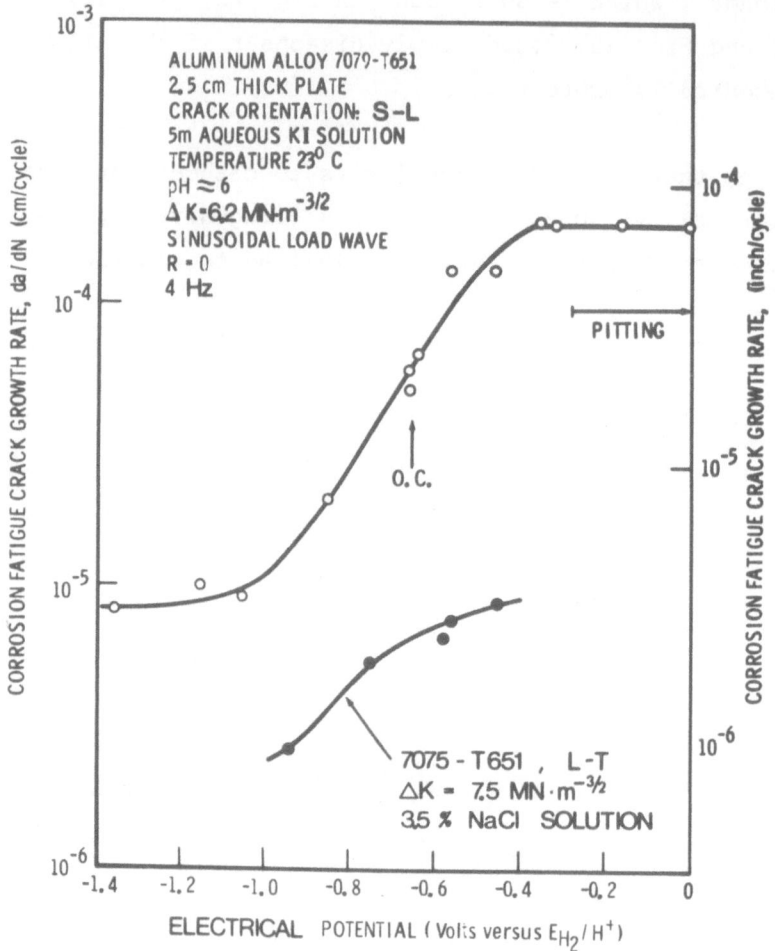

Fig. 31   Corrosion fatigue crack growth rate of two aluminum
           alloys as a function of the applied electrical
           potential[5)41)]

Inhibition of corrosion fatigue in aluminum alloys

Certain additions to water can reduce the growth rates of
corrosion fatigue cracks in aluminum alloys to below the
level normally observed in water.

The inhibitive effect of fluoride, nitrate, and chromate
as illustrated in Fig. 32, was first disclosed by the

author in 1971[43]. It is quite obvious that such additions
can reduce the growth rate of corrosion fatigue cracks
substantially, some down to the growth rates observed in
vacuum. The inhibitors mentioned are not the most practical
ones, and perhaps not the most effective ones, but the
principle was proven that corrosion fatigue crack growth in
aluminum base alloys can be retarded by inhibitors[43].
It is interesting to note that chromates are incorporated
in wet sealants in the modern aerospace industry, thus
providing an inhibitor to rivet holes where corrosion
fatigue cracks often start. The effect of fluorides mentioned
in Fig. 32 appears to shift the fatigue threshold stress
intensity ΔK_o to higher values[43]. This very significant
observation certainly merits independent double checking.
While fluorides by themselves might not be very desirable
as inhibitors, it will certainly be most desirable to search
for new inhibitors with the capability of increasing the
corrosion fatigue threshold if indeed this can be shown
to occur. Table II sums up the presently known inhibitors
for subcritical crack growth in aluminum alloys. Note that
none of them completely stops stress corrosion or corrosion
fatigue.

More recently it has been shown that nitrates inhibit
corrosion fatigue even in the presence of chlorides[42][44].
It is thought that the nitrate displaces the chloride at
the crack tip due to competitive adsorption[44]. The addition
of nitrates also causes a reversion of fatigue striation
morphology from the brittle to the ductile mode[44].

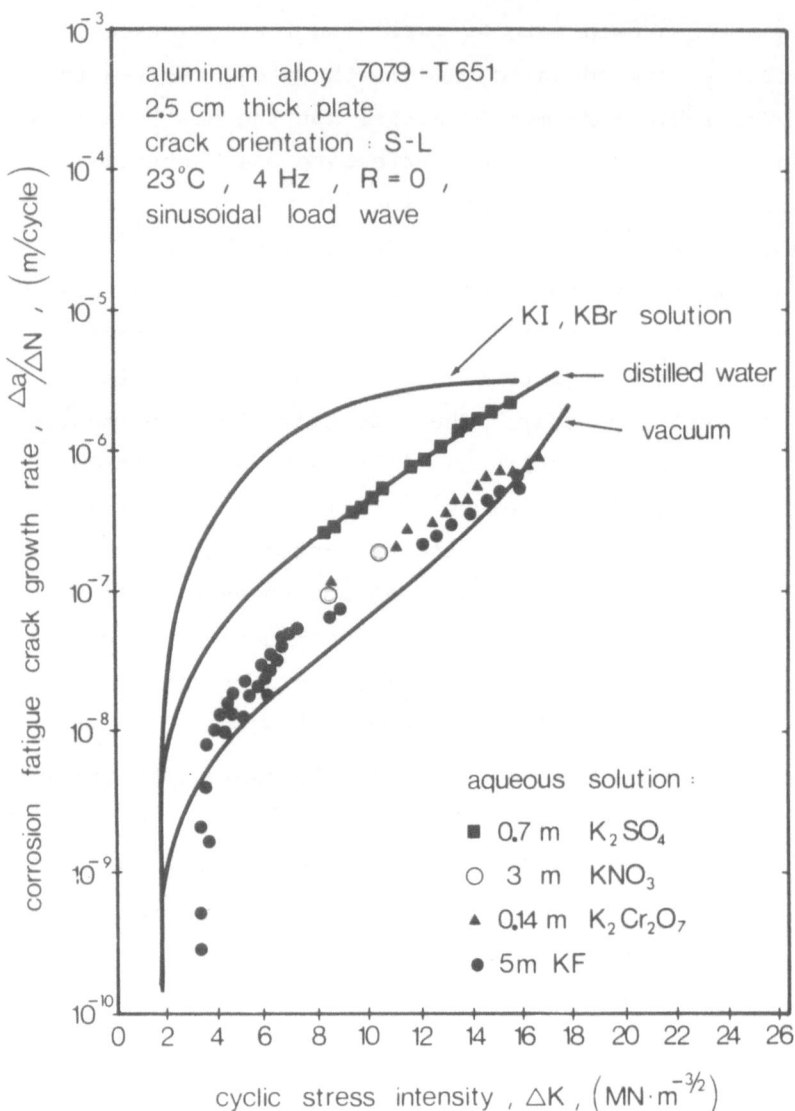

Fig. 32   Inhibiting corrosion fatigue crack growth in an
aluminum alloy

STRESS CORROSION			CORROSION FATIGUE		
Additions which <u>can</u> accelerate crack growth	Additions which do <u>not</u> accelerate crack growth	Inhibitors, retarding crack growth	Additions which <u>can</u> accelerate crack growth	Additions which do <u>not</u> accelerate crack growth	Inhibitors, retarding crack growth
Cl' Br' I'	SO_4'' CO_3'' NO_3' NO_2' CH_3COOK HCO_3'	F' Cr_2O_7''	Cl' Br' I'	SO_4''	F' Cr_2O_7'' NO_3'

Table II Accelerators, neutral additions, and inhibitors for stress corrosion and corrosion fatigue of aluminum alloys in water

References

1) Markus O. Speidel, "Current Understanding of Stress Corrosion
   Crack Growth in Aluminum Alloys", in "The Theory of Stress
   Corrosion Cracking in Alloys", J. Scully, ed. N.A.T.O.
   Scietific Affairs Division, Brussels, 1971

2) R.H. Brown, D.O. Sprowls, M.B. Shumaker, "The Resistance
   of Wrought High Strength Aluminum Alloys to Stress Corrosion
   Cracking", ASTM STP 518, (1972) pp. 87 - 118

3) Markus O. Speidel, "Stress Corrosion Cracking of Aluminum
   Alloys", Met. Trans, Vol. 6A (1975) p. 631-651

4) M.O. Speidel and M.V. Hyatt, "Stress Corrosion Cracking
   of High Strength Aluminum Alloys", in "Advances in Corrosion
   Science and Technology", Vol. 2, Plenum Press, New York, 1972

5) Markus O. Speidel et al., "Corrosion Fatigue and Stress
   Corrosion Crack Growth ....", in "Corrosion Fatigue", NACE-2,
   pp. 324-345, NACE, Houston, Texas, 1972

6) Markus O. Speidel, "Corrosion Fatigue Crack Growth in
   High-Strength Aluminum Alloys", Proceedings of 8. ICAF
   Symposium June 2, 1975, Lausanne

7) David A. Vermilyea, "A Theory for the Propagation of Stress
   Corrosion Cracks in Metals", J. Elchem. Soc. Vol. 119, (1972)
   pp. 405-407

8) John C. Scully, "Stress Corrosion Crack Propagation:
   A Constant Charge Criterion", Corrosion Science Vol. 15
   (1975) pp. 207-224

9) J.M. West, "A General Model for Stress Corrosion Crack
   Propagation", Metals Sciece Journal, Vol. 7 (1973) pp.169-171

10) C. Tyzack, "Fracture Mechanics and Stress Corrosion Cracking", Br. Corros.J. Vol. 6 (1971) pp. 219-223

11) P. Hari, "Volume Diffusion and Stress Corrosion Cracking", Corrosion Science, Vol. 15 (1975) pp 123-144

12) T.R. Beck, "Reactions and Kinetics of Newly Generated Titanium Surfaces and Relevance to Stress Corrosion Cracking", Corrosion Vol. 30 (1974) pp. 408-414

13) John Newman and William H. Smyrl, "Fluid Flow in a Propagating Crack", Met. Trans. Vol. 5 (1974) pp. 469-474

14) Tennyson Smith, "A Capillary Model for Stress Corrosion Cracking of Metals in Fluid Media", Corrosion Sciĕce, Vol. 12, (1972) pp. 45-56 and pp. 675-678

15) R.J. Gest and A.R. Troiano, "Stress Corrosion and Hydrogen Embrittlement in an Aluminum Alloy", Corrosion, Vol. 30 (1974) pp. 274-279

16) Markus O. Speidel, "Hydrogen Embrittlement of Aluminum Alloys?" in "Hydrogen in Metals", Bernstein, ed., ASM 1974, pp. 249 276

17) H.P. Van Leeuwen et al., "The Contribution of Corrosion to the Stress Corrosion Cracking of Al-Zn-Mg-Alloys", Corrosion, Vol. 31 (1975) pp. 23-29

18) H.P. Van Leeuwen, "Embrittlement by Internal and by External Hydrogen", Corrosion, Vol. 31 (1975) pp. 154-159

19) Genady P. Cherepanov, "On the Theory of Crack Growth Due to Hydrogen Embrittlement", Corrosion, Vol. 29 (1973) pp. 305-309

20) G.P.Cherepanov et al., "On the Growth of Corrosion Cracks",
    Corrosion, Vol. 29, (1973) pp. 100-104

21) L. Montgrain and P.R. Swann, "Electron Microscopy of
    Hydrogen Embrittlement in a High Purity Al-Zn-Mg Alloy"
    in Hydrogen in Metals, Bernstein, ed. ASM, 1974

22) G.T. Hahn et al., "Crack Arrest in Steel", Japan-U.S.
    Seminar on Combined Nonlinear and Linear Fracture Mechanics
    Applications 1974, Sendai, Japan

23) W.G. Clark and J.D. Landes, "An Evaluation of Rising Load
    K_{Iscc} Testing", to be published by ASTM-STP in 1976

24) R.N. Parkins et al. "Methoden zur Prüfung der Spannungs-
    korrosion", Werkstoffe und Korrosion Vol. 23 (1972)
    pp. 1020-1029

25) Markus O. Speidel, "Fatigue Crack Growth at High Temperatures"
    in "High Temperature Materials in Gas Turbines", Elsevier,
    Amsterdam, 1974

26) R.G. Forman et al., "Numerical Analysis of Crack Propagation
    in Cyclic Loaded Structures", ASME - paper Nr. 66 WA/Met. 4
    (1966)

27) T.W. Crooker, "Fatigue and Corrosion Fatigue Crack ,
    Propagation in Intermediate-Strength Aluminum Alloys"
    ASME-paper 73-Mat-N, Trans. ASME, 1974

28) Markus O. Speidel, "Stress Corrosion and Corrosion Fatigue
    in 12% Chromium Steels", in Corrosion Problems in Energy
    Conversion and Generation, Electrochemical Society,
    New York, 1974, pp. 359-367

29) F.A. McClintock, Fracture of Solids, p. 65 John Wiley,
    New York 1963

30) J.T. Staley, "Fracture Toughness and Microstructure of High Strength Aluminum Alloys", paper presented at AIME Spring Meeting Pittsburgh, May 23, 1974

31) R.P. Wei and J.D. Landes, Materials Reseach and Standard Vol. 9 (1969) p. 25

32) A.J. Mc Evily and R.P. Wei, "Fracture Mechanics and Corrosion Fatigue", in "Corrosion Fatigue", NACE, Houston, Texas, 1972, p. 381-395

33) R.P. Wei and G.W. Simmons, "Environment Enhanced Fatigue Crack Growth in High Strength Steels", Proc. Conference SCC and HE of Fe-base alloys, NACE-5, Houston, 1976

34) Markus O. Speidel, "Corrosion FAtigue in Fe-Ni-Cr Alloys", in "Stress Corrosion Cracking and Hydrogen Embrittlement of Iron Base Alloys", NACE-5, Houston, Texas, 1976

35) Markus O. Speidel, unpublished data, 1975

36) R.J. Selines and R.M. Pelloux, "Effect of Cyclic Stress Wave Form on Corrosion Fatigue Crack Propagation in Al-Zn-Mg Alloys"

37) H.P. Chu, "Fatigue Crack Propagation in a 5456-H117 Aluminum Alloy in Air and Sea Water", Report AD-769 467, 1973

38) K. Garland, "Evaluation of X 7050-T136 Die Forgings", McDonnel Aircraft Company Report 514-131.10, 1973

39) R.E. Newcomer, "Improved Aluminum Alloys", Mc Donnel Aircraft Company, Report EMA-AED-93E, 1972

40) L.R. Hall et al., "Corrosion Fatigue Crack Growth in Aircraft Structural Materials", Boeing Report AFML-TR-73-204, 1973

41) John E. Dresty and Owen F. Devereux, "The effect of specimen polarization on fatigue crack growth rates in 7075-T6 aluminum", Met. Trans. Vol. 4 (1973) pp. 2469-2471

42) R.E. Stoltz and R.M. Pelloux, "Mechanisms of Corrosion Fatigue Crack Propagation in Al-Zn-Mg", Met. Trans. Vol. 3 (1972) pp. 2433-2441

43) Markus O. Speidel, Technical Proposal, Materials and Approaches for Improved Stress Corrosion Inhibitive Coatings D 162-10423-1 to the USAF in response to RFPF 33615-71-Q-1778 January 1971, The Boeing Company Seattle, Wash. USA

44) R.E. Stoltz and R.M.N. Pelloux, "Inhibition of Corrosion Fatigue in 7075 Aluminum Alloys", Corrosion, Vol. 29 (1973) p. 13-17

THE STRESS-CORROSION CRACKING OF COPPER ALLOYS

E. N. PUGH

University of Illinois at Urbana-Champaign

INTRODUCTION

There are several types of SCC in Cu alloys, but the cracking of alpha-phase alloys in ammoniacal media is the most important practically and by far the most widely studied. This paper deals only with ammonia cracking in alpha-brass and is, in effect, an update of an earlier paper on this subject presented at Ericeira in 1971;[1] to preserve continuity, the same section headings are used in the present paper. It will be seen that the earlier view that two basically different mechanisms can operate in this system is still considered to be valid, but that a number of new developments have caused the mechanistic views put forward at Ericeira to be drastically changed.

CHARACTERISTICS OF FAILURES

It was seen in the Ericeira paper that two kinds of ammoniacal solutions exist, namely those which cause tarnishing and those which do not. SCC was reported to be either intergranular or transgranular in non-tarnishing solutions depending on the Zn content of the alloy and hence on the dislocation structure of the deformed alloy, being intergranular for Zn < ~18% (cellular dislocation structures) and transgranular for Zn >~18% (planar dislocation arrays). There had been virtually no fractographic studies of specimens failed in non-tarnishing media at that time, but recent studies[2,3] have indicated that the transgranular fractures are cleavage-like in appearance, Fig. 1, similar to those observed in a number of other systems.[4] This observation challenges the earlier view[1] that cracking occurs by a dissolution model.

Cracking in tarnishing solutions was considered to be pre-dominantly intergranular regardless of Zn content. Recent studies[3] have demonstrated that this is correct for annealed and for lightly cold worked samples, but that the path becomes transgranular in heavily cold worked specimens. For example, Fig. 2 compares the crack path for Cu-30Zn tested in a tarnishing 15 N aq NH_3 solution in (a) the annealed condition and (b) after 50% reduction by cold rolling. The transgranular fracture surfaces produced in the cold worked specimens are cleavage-like, Fig. 3(b), similar to those in the case of non tarnishing media, and it is considered that the mechanism is the same. Thus a competition is envisaged in tarnish-ing solutions between the cleavage-like and intergranular fracture modes, and the former is thought to predominate in heavily cold worked material.

The fracture surfaces in the cold worked material exhibited a series of parallel markings, Fig. 3(b), similar in appearance to the "striations" reported by McEvily and Bond[5] for Cu-30Zn tested in Mattsson's tarnishing solution of pH 7.2. McEvily and Bond sug-gested that the markings are similar to fatigue striations and that they indicate that crack propagation is discontinuous.

Mechanistically, this is an extremely important claim, and therefore it should be emphasized that the detailed fractographic studies
necessary to establish its validity have not been carried out.
McEvily and Bond reported parallel markings for annealed and heavily
cold worked material, and, because of the use of the tarnishing
environment, it has been <u>assumed</u> that cracking was intergranular
in both cases. However, in the light of the preceding paragraph it
must now be recognized that the path of cracking in the cold worked
case was almost certainly transgranular. Similarly, the fracture
surfaces studied by Pugh, Craig and Sedriks[6] in cold-worked Cu-30Zn
were probably transgranular, although the authors assumed them to
be intergranular. The question now arises whether the markings have
the same origin on inter- and transgranular surfaces. Studies of
the former have frequently led to the observation of parallel markings, e.g. Fig. 3(a), but in all cases these have been shown to
correspond to slip steps.[3,7-9] The markings on the transgranular
fracture surfaces, e.g. Fig.3(b),may prove to be striations but, as
discussed elsewhere at the Institute,[4] the necessary fractographic
studies have not yet been completed.

A feature emphasized in the Ericeira paper was the preferential
penetration of grain boundaries in unstressed alpha-brasses by the
tarnish, Fig. 4. It has been pointed out that a correlation exists
between the occurrence of intergranular attack and intergranular
SCC, and this has led to the view that this form of SCC occurs by
stress-assisted intergranular attack.[1,10,11] However, recent observations are presented in following sections which seriously
weaken this model.

The strong dependence of the stress-corrosion susceptibility
of the alpha-brasses on zinc content is well established but the

behavior of unalloyed copper remains controversial. The question
of the susceptibility of copper in non-tarnishing ammoniacal solutions had not been resolved at Ericeira and remains in doubt. At
that time, there was general agreement that the unalloyed metal
is immune to cracking in tarnishing media. Subsequent work[12] has
demonstrated that intergranular cracking, similar to intergranular
stress-corrosion cracks, can be produced in slow strain-rate tensile
tests in which copper is immersed in tarnishing solutions, Fig. 5.
The significance of this observation is discussed in the section on
mechanisms.

CHEMISTRY OF THE SYSTEM

Understanding of the behavior of copper and alpha-brass in
non-tarnishing solutions remains essentially unchanged since the
Ericeira paper.[1] The system is under concentration polarization,
anodic dissolution of copper and zinc proceeding at a rate determined by the transport of the predominant cathodic species,
$Cu(NH_3)_n^{2+}$, to the surface. No protective films exist in the non-
tarnishing range,* and large rates of active anodic dissolution
are possible, producing faceted surfaces. There is no preferential
dissolution of zinc on a macroscopic scale, but it is possible that
a shallow dezincified layer exists. Auger analysis by Pinchback,
Clough and Heldt[2] provided evidence for dezincification at transgranular stress-corrosion fracture surfaces produced in a non-tarnishing 15 N solution, but no studies have been made of unstressed
samples.

---

*Brown loosely adherent films are commonly observed in the non-
tarnishing range, but these afford no protection.[13,14] The
origin of these films has not been established.

Tarnishing occurs when a critical concentration of $Cu(NH_3)_n^{2+}$ ions is attained under open-circuit conditions,[1] or when a critical anodic current is reached in potentiostatic tests.[12] There have been a number of advances in our understanding of the nature of the tarnish. In the Ericeira paper, the tarnish was considered to consist of a single layer of Cu_2O, formed by a single-stage process. More recent studies have indicated that tarnishing probably involves several steps:

(i)   the formation of a thin protective film of Cu_2O.

(ii) local breakdown of the thin film, leading to rapid anodic dissolution of the substrate and the precipitation of a thick porous layer of Cu_2O.

(iii) the formation of an outer layer of CuO.

Steps (i) and (ii) were suggested earlier by Jenkins and Durham on the basis of optical microscopic studies of copper and Cu-30Zn exposed to 15 N aq NH_3, but the evidence for the existence of a protective film was rather meager. However, recent studies by Cheng[12] using 1 N solutions leave little doubt that such a film is formed. For example, the thickness of the surface film was found to attain a limiting value of ~200Å, Fig. 6, indicating its protective nature. Cheng also observed an active-passive transition at the onset of tarnishing in anodic-polarization experiments, Fig. 7; the increasing currents at larger anodic potentials are thought to result from film breakdown. A significant feature of Cheng's work was the observation that the growth kinetics were not strongly dependent on zinc content, see Fig. 6, but that the rate of film breakdown increased with increasing zinc content.

The ellipsometric data of Fig. 6 are in marked contrast to those reported earlier by Green, Mengelberg and Yolken.[15] Thus, using a

15 N solution containing 8 g/ℓ dissolved copper, Green et al. observed (i) extremely rapid growth kinetics, e.g., the film on Cu-30Zn attained a thickness of 900Å in ~1 sec; (ii) the kinetics increased dramatically with increasing zinc content, e.g., the time for the film to grow to 900Å was 44 min for pure copper, and (iii) the kinetics were linear for copper in the range studied, 0-3000Å. The differences between the 1 N and 15 N solutions are thought to be due to the rapid breakdown of the protective film in the latter, so that the film studied by Green et al. corresponded to the precipitated porous layer, whereas Cheng's data for 1 N solutions relate to the protective film. Work is in progress to confirm this view.

The most significant recent observations on the tarnishing process and its relation to SCC were made by Pinchback, Clough and Heldt,[9] who studied intergranular stress-corrosion fracture surfaces produced in Cu-30Zn tested in a 15 N solution containing 8 g/ℓ Cu, added as cupric hydroxide. Specimens containing intergranular stress-corrosion cracks were removed from the solution, rapidly bent in air to produce complete fracture, and the fracture surfaces first examined with the optical microscope. A color transition was observed, the ductile region being brass colored while the stress-corrosion fracture surface varied from copper colored, dark yellow, mixed black and yellow, and black with increasing distance from the crack tip; the black region was characteristic of the thick tarnish on the external surfaces of the specimens. Using the SEM, the stress-corrosion fracture surface near the crack tip appeared smooth and film free, and displayed occasional parallel markings considered to be slip steps, Fig. 8(a). Crystalline deposits or platelets were observed at distances of several grains from the tip, Fig. 8(b), and their density increased with increasing distance until they grew together to form a continuous film, Fig. 8(c), which extended to the free surface.

Pinchback et al. then carried out chemical analysis of the fracture surfaces using energy dispersive X-ray analysis (EDAX) and Auger electron spectroscopy (AES). Typical results are summarized in Fig. 9, in which the ratio of the atomic concentrations of copper and zinc are plotted versus distance from the crack tip. The EDAX measurements indicated a large increase in the Cu/Zn ratio in the region covered by the continuous film. The ratio decreased in the region of crystalline deposits to about the value for the bulk alloy and this persisted up to the crack tip and across the ductile fracture surface. The AES data also showed a large Cu/Zn ratio in the region covered by the continuous film. This value decreased as the crack tip was approached, but remained _above_ that for the bulk alloy until the ductile fracture surface was reached.

From these observations, Pinchback et al. concluded that an oxide film exists on the stress-corrosion fracture surface near the crack tip, the film being thick enough for AES detection but too thin for the sensitivity of EDAX. The crystalline deposits at greater distances from the crack tip were considered to result from breakdown on the thin oxide film, in accord with the Jenkins and Durham model discussed above, and the continuous film to correspond to the thick precipitated layer. In effect, the observations illustrate steps (i) and (ii) above, and thus represent an important contribution to our understanding of the tarnish-growth process. In addition, they have direct relevance to the mechanism of SCC, since they indicate that the protective film extends to the crack tip and that its breakdown occurs at some distance behind the

advancing crack front. The significance of this observation will
be discussed in the section on mechanisms, but it is apparent that
attention must now be focussed on the protective or passive film
rather than on the thick outer layer, the latter playing no direct
role in the stress-corrosion process.

The preceding discussion implies that the tarnishing behavior
of the stress-corrosion fracture surfaces is essentially the same
as that of bulk alloy. However, several observations suggest that
this may be an oversimplification. For example, Tromans, Dowds and
Leja,[7] studying specimens partially failed by SCC and then fractured
rapidly in the tarnishing solution, showed that the stress-corrosion
surfaces were yellow while the ductile region was covered with the
black tarnish; etching with 45% HNO_3 and re-immersion in the test
solution caused the fracture surfaces to tarnish uniformly. Based
on more recent observations,[8,9] these results suggest that the
kinetics of tarnishing are slower on the stress-corrosion fracture
surfaces. Recently, Procter and Stevens[16] have pursued this possibility. In addition to confirming the observations of Tromans
et al.,[7] they showed that on chemical or electrochemical removal of
the cuprous oxide film from the stress-corrosion fracture surface
by techniques which do not involve dissolution of the substrate,
the fracture surface does not "tarnish" on re-immersion in Mattsson's
solution of pH 7.2. Further, cathodic-reduction experiments indicated
that the rate of tarnish formation was very much smaller than that
of ductile fracture surfaces or metallographically polished surfaces.
Procter and Stevens suggested that surfaces produced by SCC are
resistant to tarnishing, and that this results from dezincification
at the crack tip during cracking. Such a view is consistent with
the observation of Cheng[12] that the rate of breakdown of the protec-

tive film increases with increasing zinc content, and with other experimental evidence which suggests that intergranular SCC in alpha-
brass involves dezincification.[17,18]

There have been no systematic investigations of the structure
and composition of the protective film, or of the breakdown process.
Diffraction studies leave little doubt that it consists of Cu_2O,[8,14]
and the Auger data of Pinchback et al.[9] indicate that in the case of
Cu-30Zn it contains substantial amounts of zinc, Fig. 9. From
ellipsometric and electrical-conductivity measurements, Roberts,
Booker, Osborne and Salim,[19] also concluded that the tarnish films
formed on copper and a range of Cu-Zn alloys in Mattsson's solution
of pH 7.2 contain large amounts of zinc, so that the films contain
large concentrations of holes and cation vacancies. For example,
they observed that the conductivities of the films were about 10 or
100 times greater than that for Cu_2O formed by thermal oxidation.
Unfortunately, Roberts et al. were unaware that tarnishing involves
two steps, and their data are now ambiguous because it is not clear
whether they were dealing with the protective film, the thick porous
layer, or mixtures of the two. With respect to SCC, it will be seen
below that the details of the re-passivation process now become of
prime importance, particularly concerning the effect of zinc, but
no data currently exist in the literature.

There is now considerable information on the outer layer. This
can grow to considerable thicknesses (>5 um; see Fig. 4), and
its composition can be readily determined by means of the electron
microprobe. Such studies[20,21] have established that the thick
layers in alpha-phase Cu-Zn, Cu-Ni and Cu-Al alloys are essentially
depleted with respect to the alloying elements, e.g. Fig. 10.
These observations strengthen the view that the layer is formed by

a dissolution-reprecipitation mechanism. Thus it is considered that both copper and the alloying elements enter solution at the interface between the metal and the porous tarnish, and that the alloying elements remain in solution as ions, diffusing through the pores to the bulk solution, while the copper ions are precipitated as Cu_2O. There is no direct evidence for the existence of pores, and the view that the layer is porous is based on the rapid growth kinetics observed by Green et al.[15]

Prolonged exposure to the tarnishing solution leads to the formation of an outer layer, Fig. 10(a), shown by X-ray diffraction to be CuO.[21] The thickness of this layer increases with time of immersion, so that eventually only CuO is detected in diffraction experiments,[21] and it is possible that this accounts for the apparently anomalous observation by Sparkes and Scully[22] that the tarnish formed on Cu-1.8Be consisted of CuO. The mechanism of formation of the CuO layer has not been established, but it may result from the oxidation of Cu_2O. It is interesting to note that the corrosion potential undergoes a noble shift during tarnish growth, and that the initial and final values are in the regions of stability of Cu_2O and CuO, respectively, on the Pourbaix diagram.[12]

An interesting feature of the microprobe data of Gabel et al.[21] is that the copper contents of the oxides were approximately 5% below the stoichiometric values of 88.8 and 79.9 wt %, e.g. Fig. 10(b). It is possible that the porous nature of the oxides introduces error in the microprobe results, although the literature suggests that this is unlikely. If the departure from stoichiometry is real, then the oxides would be highly defected and might be expected to display unusual electrical properties. It is attractive to speculate that the large conductivities reported by Roberts et al.[19] result from the departure from stoichiometry. However, as noted above, it is not clear whether Roberts et al, were dealing with the protective film or the thick layer.

MECHANISM OF FAILURE

A number of failures must be considered:

(i)   Intergranular SCC in non-tarnishing solutions (in alloys with <~18%Zn).

(ii) Transgranular SCC in non-tarnishing solutions (in alloys with >~18%Zn).

(iii) Intergranular SCC in tarnishing solutions (in all alloys in the annealed condition).

(iv) Transgranular SCC in tarnishing solutions (in heavily cold worked Cu-30Zn).

It is not known if a single mechanism is responsible for all of these failures or if two or more basically different mechanism exist. At Ericeira, it was suggested that (i) and (ii) occur by the same mechanism and that this is basically different from that operative in (iii). This view is still considered valid although it will be seen that the details of the respective models have been drastically changed. Failure (iv) was not recognized at Ericeira; it will be argued that it is essentially the same as (ii).

Mechanism in Tarnishing Solutions

This section deals with case (iii), the intergranular SCC of annealed alpha-brasses in tarnishing solutions; the failure probably corresponds to the practical failure season-cracking. In the Ericeira paper, it was suggested that it occurs by the repeated formation and rupture of the thick tarnish film, which was thought to grow preferentially into the metal along grain boundaries. It was seen above that the observations of Pinchback et al.[9] provide convincing evidence that the thick film does not extend to the crack tip, so that it is clear that the tarnish-rupture model is untenable in its present form. The model could be modified by simply replacing the thick film by the protective film, which is in fact present at the crack tip.[9] It should be noted, however, that no evidence exists for intergranular penetration by the protective film. The occurrence of striated fracture surfaces was also taken as evidence for the tarnish-rupture model, but it was seen above that there is in fact no evidence for the presence of striations on the intergranular surfaces.

The recognition that a protective film exists at the crack tip suggests alternatively that the classical Champion-Logan film-rupture model[23,24] may be operative. This model is based on the concept that plastic deformation exposes film-free metal at the crack tip, permitting crack propagation to occur by localized anodic dissolution. The details of the film-rupture model have been discussed by several workers in recent years, e.g. [11,25,26] For the present purpose, it should be noted that a competition is envisioned between anodic dissolution and repassivation at the crack tip. At Ericeira, for example, Engell[25] stated that "the potential must be high enough to allow rapid anodic dissolution at the crack tip as well as the formation of a surface layer apart from the tip, but should not be so high that passivation of the crack tip occurs." Such a condition can be fulfilled only by certain environments, accounting for the specific nature of solutions which cause SCC. This argument may explain the observation that intergranular cracking of brass occurs only in aq NH_3 or certain aq citrate and tartrate solutions.[1]

Many of the other characteristics of the intergranular SCC of Cu-Zn in tarnishing aq NH_3 can also be rationalized in terms of the film-rupture model. The requirement that propagation takes place only by anodic dissolution sets an upper limit for the crack velocities which can be expected, and this rules out the model for certain systems.[25,27] Surprisingly, there is little data in the literature on the rates of intergranular SCC in brasses. The values reported by McEvily and Bond[5] for Cu-30Zn (10^{-5}-10^{-4} cm/s) were obtained for heavily cold worked material and it was seen above that cracking was therefore probably transgranular. Heinzel, Rothenbacher and Engell[28] reported crack propagation rates between 10^{-5} and 10^{-4} cm/s for a range of Cu-Zn alloys, but in this case the solution was non-tarnishing (15N aq NH_3 containing 0.125 mole/l $Cu(NO_3)_2$), cracking being transgranular in Cu-30Zn. An estimate of the average velocity of intergranular SCC can be obtained from published time-to-failure data,[29] presented at Ericerira.[1] Assuming no incubation period, the average velocity for Cu-30Zn tested in a tarnishing 15N solution containing 8g/l copper is ~0.25 x 10^{-3} cm/s, and, from Faraday's law,[25,27] this would require an anodic current density at the crack tip of ~3.4A/cm^2. This

value is considerably larger than the maximum observed anodic current densities for the unstressed alloy in non-tarnishing 15N
solutions, viz., $\sim 0.02/cm^2$ for unstirred and $\sim 0.1A/cm^2$ for stirred
solutions.[1,30] However, the latter are limited by transport of the
cathodic species to the surface, whereas this would not be expected
to be the case for the film-free metal at the top of an advancing
crack, where a large cathodic area exists, i.e., the faces of the
crack. Engell[25] has pointed out the the maximum current densities
observed for the anodic dissolution of metals ranges from $10A/cm^2$
for aluminum to $100A/cm^2$ for iron. Thus the value of $\sim 3.4A/cm^2$
does not appear to be prohibitively large.

An important characteristic of intergranular SCC in alpha-
brasses is the marked dependence of susceptibility on zinc content.
For example, data presented in the Ericeira paper[1,29] for a tarnishing 15N solution containing 8g/l copper indicated that the times-to-
failure, t_f, were 3400, 350, 200 and 90 s for specimens containing
5, 10, 20 and 30% zinc, respectively. In terms of the film-rupture
model, this effect may be attributed to the effect of zinc on two
related processes, namely the anodic dissolution of the film-free
metal and repassivation. Studies of the anodic kinetics for Cu-Zn
alloys in non-tarnishing 15N solution indicated that the zinc content has little effect on the Tafel slopes but that the total anodic
current increased by a factor of ~ 6.5 from 10 to 30% zinc,[12] cf.,
with a factor of ~ 4 for t_f. No data exists in the literature on
the kinetics of repassivation in this system, and this is clearly
an area requiring attention. Another factor which may influence

susceptibility is slip mode, since large areas of film-free surface would be produced at slip steps in alloys displaying planar slip (Zn > 15%) compared to those in which cellular dislocation structures occur (Zn < 15%).

The behavior of pure copper in tarnishing solutions can also be rationalized in terms of the film-rupture model if it is assumed that the repassivation rate increases with decreasing zinc content. A competition is envisioned between anodic dissolution and repassivation at the crack tip, and thus it is possible that the latter predominates in the unalloyed metal. In constant load tests, the metal would therefore be immune, but in slow strain-rate tests clean surface would continually be created at the crack tip, accounting for the observed cracking, Fig. 5. Parkins[31] has recently discussed the role of strain rate in the film-rupture model, pointing out that "if crack propagation occurs by dissolution at an essentially film-free crack tip, with the crack sides rendered inactive by filming, the maintenance of film-free conditions may be dependent not only upon the electrochemical conditions but also upon the rate at which bare metal is created at the crack tip by plastic strain." He pointed out that a maximum degree of cracking can be expected at an intermediate rate in slow strain-rate tests, ductile failure occurring at high rates when the ductile crack can propagate faster than the stress-corrosion crack and also at low rates when the repassivation rate exceeds that of creation of new surface. In constant load tests, he suggests that crack propagation will continue only when the strain rate at the crack tip (creep) exceeds the minimum rate. No systematic study has been made of the effect of strain rate on the cracking of pure copper, but it is evident that such studies would be valuable.

A final point to be discussed is the intergranular path of cracking. It was once considered that the film-rupture model is relevant only to transgranular cases of SCC,[32] but it was later recognized that the structure and composition of the grain boundaries might cause the repassivation kinetics to be slower at the boundaries than at the grain surfaces,[26,33] leading to an intergranular path of cracking. The penetration of boundaries in unstressed specimens by the thick tarnish, Fig. 4, may also prove to result from the slower repassivation kinetics at the boundaries. Thus, rather than being the basic cause of crack propagation, as suggested in the Ericeira paper, intergranular penetration is now considered to be a secondary effect, related to the underlying cause -- the slower repassivation kinetics at grain boundaries. It should be noted, however, that the preceding discussion has not considered crack initiation, and it is possible that intergranular penetration plays an important role in this process.

Mechanism in Non-Tarnishing Solutions

Cracking in non-tarnishing solutions occurs when stressed samples undergo rapid anodic dissolution; in open-circuit tests, the presence of cupric complex ions is necessary to provide sufficiently rapid dissolution rates, but cracking can be produced in the absence of the complexes in potentiostatic tests.[1,30] In the Ericeria paper, it was proposed that the intergranular and transgranular cracking in non-tarnishing solutions occur by the same mechanism, and that the failures involve preferential anodic dissolution at the crack tip. The absence of protective surface films ruled out the film-rupture model, and it was suggested that failure involved preferential dissolution at dislocations. Since that time, SEM studies have shown that the transgranular fracture surfaces in Cu-30Zn are cleavage-like in appearance, Fig. 1, and, as discussed elsewhere at the Institute,[4] it is difficult to reconcile such fracture surfaces with a dissolution model.

If one accepts the view that the transgranular failures occur by brittle mechanical fracture, the question then becomes one of determining the cause of embrittlement. Several possibilities exist:

(i) The adsorption model, in which it is proposed that specific species adsorb and interact with strained bonds at the crack tip, causing a reduction in bond strength and permitting brittle fracture.[34,35] The identity of adsorbing species has not been established. Uhlig, Gupta and Liang[35] have suggested that they are copper-containing or other heavy-metal ammonium complexes, but there is no unambiguous evidence for this. It was seen above that cupric complex ions can be ruled out[1,30], but it is possible that the cuprous species are responsible. Alternatively, it might be argued that the ammonia molecules or the ammonium ions are the critical species, and that the role of the cupric complex is simply to increase the corrosion potential above the postulated critical potential for SCC.[34,35] Similarly, anodic polarization would be expected to cause cracking.

(ii) By analogy with other stress-corrosion failures which exhibit cleavage-like fracture surfaces,[4] it might be suggested that the transgranular fractures are due to hydrogen embrittlement. However, the source of the hydrogen is not evident. Since the rate of cracking increases as the rate of anodic dissolution increases,[1,30] it might be speculated that the anodic reaction liberates hydrogen, e.g.

$$Cu + 2NH_4^+ = Cu(NH_3)_2^+ + 2H^+ + e$$

There is no evidence of hydrogen evolution during dissolution or cracking, however; nor does exposure of unstressed specimens lead to embrittlement in subsequent tensile tests.[3]

(iii) In an early paper, Forty[36] suggested that dezincification might produce an embrittled zone at the crack tip, leading to discontinuous brittle fracture. This possibility has received little attention, but it was noted above that Pinchback et. al.[2] have provided evidence for dezincification at transgranular stress-corrosion fracture surfaces.

Similar cleavage-like fractures were observed in heavily cold worked Cu-30Zn tested in tarnishing solutions, Figure 3(b), and it seems likely that the mechanism is the same. It was seen that the fracture surfaces in the cold worked material exhibit parallel markings which may prove to be striations. Such markings have not been reported in the non-tarnishing case, but this may simply be due to the fact that such surfaces have received little attention. If failure is in fact due to cleavage, then propagation would be expected to be discontinuous, since cleavage cracks propagate at velocities approaching that of sound waves, whereas the overall rate of transgranular cracking was seen to be relatively small, $\sim 10^{-4}$ cm/s.[5,28] A discontinuous mode of cracking is consistent with (ii) and (iii), but not with (i) since the adsorption model would appear to predict continuous cracking at a rate controlled by transport of the critical species to the crack tip.

So far in this section, discussion has been confined to the transgranular cracking which occurs in the low SFE alloys. It is not known if the intergranular cracking in alloys with Zn < ~ 18% occurs by the same mechanism. If the mechanism were common then the change in crack path would presumably be attributed to the change in dislocation behavior. It is clearly premature to speculate further about the mechanism in the intergranular case, but two points can be made. First, the occurrence of SCC in unalloyed copper[1] rules out the dezincification model as a generalized mechanism. Second, the path of cracking can be intergranular in the hydrogen embrittlement of certain materials with high SFE, e.g., Ni and high Ni alloys.[37,38]

CONCLUSIONS

The most significant change since the Ericeira paper concerns the intergranular SCC of annealed alpha-brasses in tarnishing solutions. The recognition that the tarnishing process consists of two steps, the formation of a thin protective film and its subsequent breakdown to form a thick precipitated layer, and that it is the thin film which exists at the tip of the intergranular cracks, has caused the tarnish-rupture model to be discarded. It is suggested that the recent observations can be accounted for by the film-rupture model, that is, by preferential anodic dissolution of film-free metal at the crack tip where plastic deformation continually ruptures the protective film. The penetration of grain boundaries in unstressed specimens by the thick layer, previously regarded as strong evidence for the tarnish-rupture model, is now thought to be simply a secondary manifestation of the underlying cause of cracking--the slower repassivation kinetics at the grain boundaries.

The observation that the transgranular fracture surfaces of alloys containing >18% Zn tested in non-tarnishing solutions are cleavage-like in appearance has also caused the mechanistic viewpoint to be revised in this case. Rather than a dissolution model, it is now suggested that failure occurs by brittle mechanical fracture, probably discontinuous in nature. The basic cause of embrittlement is not understood, but both dezincification and hydrogen embrittlement are recognized as candidates. Similar cleavage-like failures

have been observed in heavily cold-worked Cu-30Zn in tarnishing
solutions, and it is thought that the same mechanism is operative.
Thus, in tarnishing solutions it appears that two competing
mechanisms exist, the intergranular one predominating in annealed
or lightly worked material but the transgranular mode occurring in
the high-zinc alloys when they are in the heavily cold-worked
condition.

The intergranular SCC of annealed alloys with <18% Zn in non-
tarnishing solutions has received little attention, and the mechanism
is not understood. It is possible that the mechanism is the same
as that in the transgranular case, and the the change in the path
of cracking is due to the change in dislocation structure. Note
that the absence of protective surface films is considered to rule
out the film-rupture model in this case.

Thus it is concluded that at least two different mechanisms
can operate in the alpha brass-aq NH_3 system, the film-rupture
mechanism in the case of intergranular failures in tarnishing
solutions, and a model involving brittle mechanical failure
in transgranular cases. In the Ericeira paper, it was stated
that the latter has no commercial importance, but since that
time the author has become aware that this mode of failure
is generally responsible for commercial failures in Admiralty
Metal condenser tubes, a problem of considerable practical
relevance.

ACKNOWLEDGMENTS

The author is again indebted to H. M. Davis for his
encouragement and for his numerous helpful suggestions, and to
the U.S. Army Research Office for financial support.

REFERENCES

1. E. N. PUGH, "The Theory of Stress Corrosion Cracking in Alloys" ed. J. C. SCULLY), p. 418, NATO Scientific Affairs Div., Brussels (1971).

2. T. R. PINCHBACK, S. P. Clough, and L. A. HELDT, Corrosion in press.

3. J. A. BEAVERS, H. S. TONG, and E. N. PUGH, work supported by U.S. Army Research Office under grant ARO-D-127. To be published.

4. E. N. PUGH and D. G. CHAKRAPANI, this volume.

5. A. J. McEVILY, JR., and A. P. BOND, J. electrochem. Soc., 112, 131 (1965)

6. E. N. PUGH, J. V. CRAIG, and A. J. SEDRIKS, "Fundamental Aspects of Stress-Corrosion Cracking" (ed. R. W. STAEHLE, A. J. FORTY, and D. VAN ROOYEN), p. 118, NACE, Houston (1969).

7. D. TROMANS, N. A. DOWDS, and J. LEJA, ibid., p. 154.

8. S. S. BIRLEY and D. TROMANS, Corrosion, 27, 297 (1971).

9. T. R. PINCHBACK, S. P. CLOUGH, and L. A. HELDT. Met. Trans., 6A, 1479 (1975)

10. J. A. BEAVERS, L. C. ROSENBERG, and E. N. PUGH, "1972 Tri Service Conference on Corrosion" (ed. M. M. JACOBSON and A. GALLACCIO), p. 57, MCIC Report 73-19, Battelle (1973).

11. E. N. PUGH, "Proc. Conf. on Stress-Corrosion Cracking and Hydrogen Embrittlement of Iron Base Alloys" (ed. R. W. STAEHLE), NACE, in press.

12. B. C. CHENG, Ph.D. Thesis, University of Illinois at Urbana-Champaign, 1975.

13. E. N. PUGH and A.R.C. WESTWOOD, Phil. Mag., 13, 167 (1966).

14. L. H. JENKINS and R. B. DURHAM, J. electrochem. Soc., 117, 768 (1970).

15. J.A.S. GREEN, H. D. MENGELBERG, and H. T. YOLKEN, ibid, p.433

16. R.P.M. PROCTER and G. N. STEVENS, Corros.Sci. 15, 349, (1975)

17. H. LEIDHEISER and R. KISSINGER, Corrosion, _28_, 218 (1972).

18. H. W. PICKERING and P. J. BYRNE, Corrosion, _29_, 325 (1973).

19. E.F.I. ROBERTS, C.J.L. BOOKER, P. OSBORNE, and M. SALIM, Corros. Sci., _14_, 307 (1974).

20. A. J. FORTY and P. HUMBLE, "Environment-Sensitive Mechanical Behavior (ed. A.R.C. WESTWOOD and N. S. STOLOFF), p. 403, Gordon and Breach, New York (1966).

21. H. GABEL, J. A. BEAVERS, J. B. WOODHOUSE, and E. N. PUGH, Corrosion, in press.

22. G. M. SPARKES and J. C. SCULLY, Corros. Sci., _11_, 641 (1971).

23. F. A. CHAMPION, "Symposium on Internal Stresses in Metals and Alloys," p. 468, Institute of Metals, London (1948).

24. H. L. LOGAN, J. Res. Natn. Bur. Stand., _48_, 99 (1952).

25. H. J. ENGELL, "The Theory of Stress Corrosion Cracking in Alloys" (ed. J. C. SCULLY), p. 86, NATO Scientific Affairs Div., Brussels (1971).

26. R. W. STAEHLE, ibid., p. 223.

27. E. N. PUGH, J.A.S. GREEN, and A. J. SEDRIKS, "Interfaces Conference - Melbourne 1969," p. 237, Butterworth (1969).

28. H. HEINZEL, P. ROTHENBACHER, and H. J. ENGELL, Z. Metallkunde, _61_, 511 (1970).

29. E. N. PUGH, J. V. CRAIG, and W. G. MONTAGUE, Trans. ASM, _61_, 468 (1968).

30. E. N. PUGH and J.A.S. GREEN, Met. Trans., _2_, 3129 (1971).

31. R. N. PARKINS, "Proc. Conf. on Stress-Corrosion Cracking and Hydrogen Embrittlement of Iron-Base Alloys" (ed. R. W. STAEHLE), NACE, in press.

32. R. N. PARKINS, Met. Rev., _9_, 201 (1964).

33. A. J. SEDRIKS, P. W. SLATTERY, and E. N. PUGH, "Fundamental Aspects of Stress-Corrosion Cracking" (ed. R. W. STAEHLE, A. J. FORTY, and D. VAN ROOYEN), p. 673, NACE, Houston (1969).

34. H. H. UHLIG, "Proc. Conf. on Stress-Corrosion Cracking and Hydrogen Embrittlement of Iron-Base Alloys," (ed. R. W. STAEHLE), NACE, in press.

35. H. H. UHLIG, K. GUPTA and W. LIANG, J. electrochem. Soc.,
     122, 343 (1975).

36. A. J. FORTY, "Physical Metallurgy of Stress-Corrosion
     Fracture," (ed. T. N. RHODIN), p. 99, Interscience,
     New York (1959).

37. M. L. WAYMAN and G. C. SMITH, Met. Trans., 1, 1189 (1970).

38. M. L. WAYMAN and G. C. SMITH, Acta Met., 19, 227 (1971).

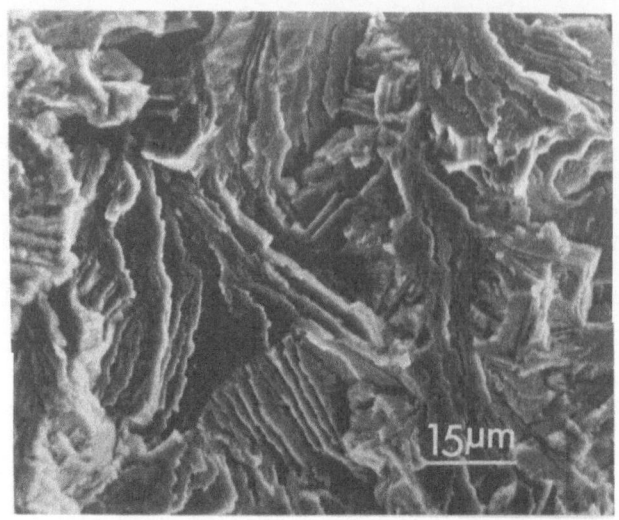

Fig. 1.   Scanning electron micrograph of stress-corrosion fracture
surface (transgranular) for a Cu-30Zn specimen tested in a
non-tarnishing 15N ammoniacal solution.[2]

Fig. 2. Scanning electron micrographs of the surfaces of Cu-30Zn
stress-corrosion specimens tested in a 15N tarnishing
solution. (a) annealed (intergranular); (b) 50% reduction by cold work (transgranular).[3]

Fig. 3. Scanning electron micrographs of fracture surfaces of
specimens shown in Fig. 2. (a) annealed -- intergranular
showing slip steps; (b) 50% reduction by cold work --
transgranular, cleavage-like surface.[3]

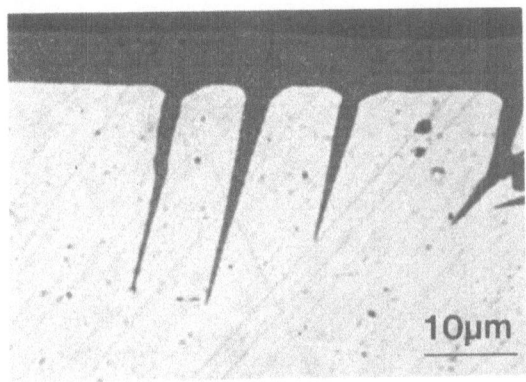

Fig. 4.   Optical micrograph of section of Cu-30Zn sheet which was
immersed (unstressed) for 5 days in a tarnishing 15N
solution, illustrating preferential penetration of the
tarnish along grain and twin boundaries.[10]

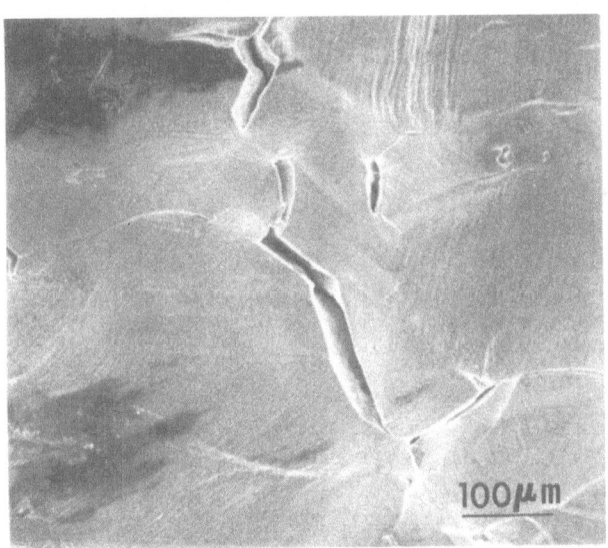

Fig. 5.   Scanning electron micrograph showing intergranular
cracks at the surface of a copper tensile specimen
tested at a strain rate of 0.002 min^{-1} in a tarnishing
1N solution.[12]

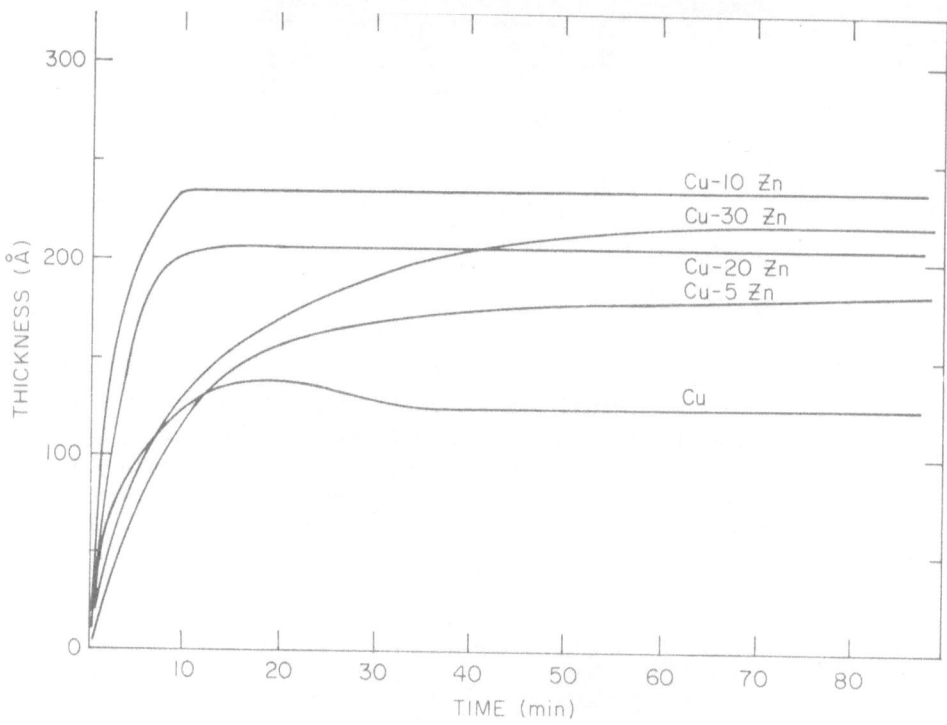

Fig. 6. Variation of film thickness, determined ellipsometrically with time for specimens immersed in 1N aq NH_3 containing 1 g/l copper dust.[12]

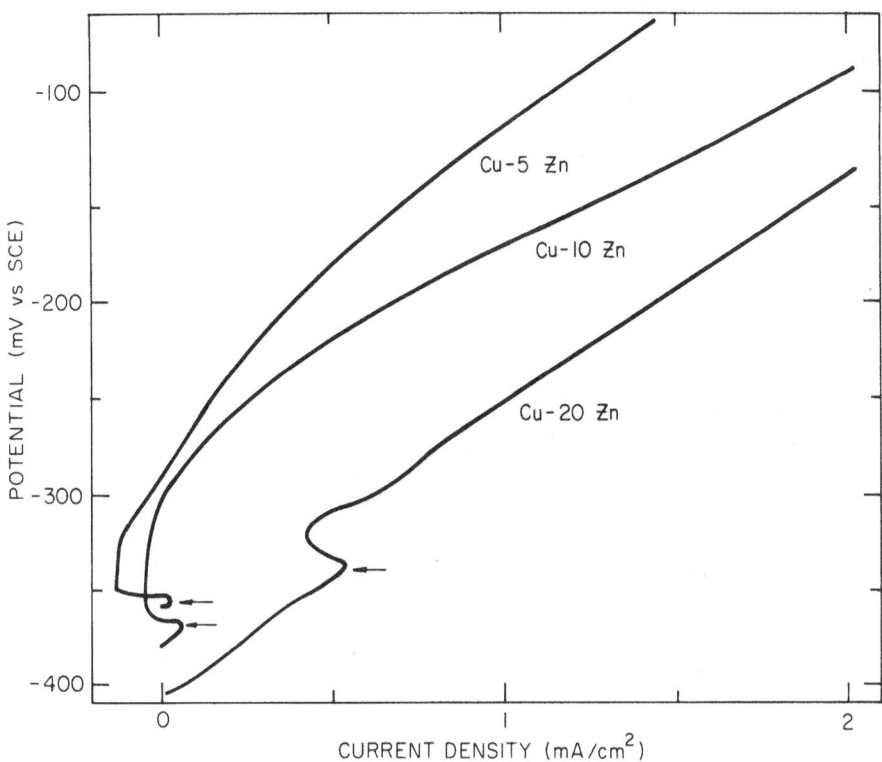

Fig. 7. Anodic polarization curves for specimens in 1N aq NH_3 containing 0.5 g/l copper dust. The arrows indicate the onset of tarnishing in each case.[12]

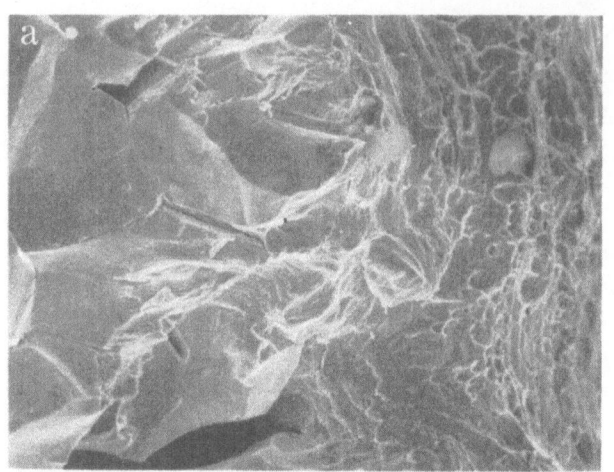

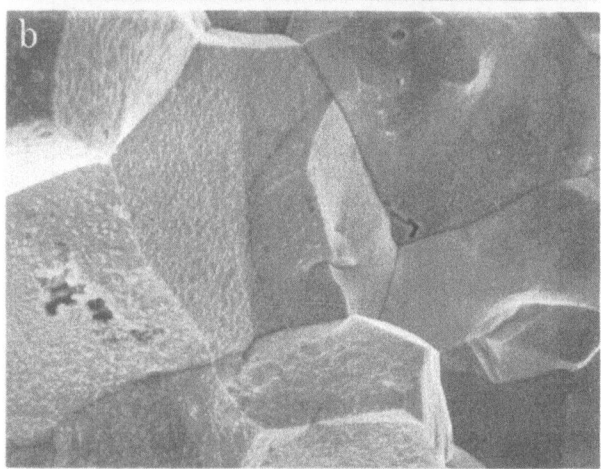

Fig. 8. Scanning electron
micrographs of intergranular
stress-corrosion fracture
surface of Cu-30Zn tested in
a tarnishing 15N solution.
(a) illustrates a region at
the tip of the stress-corrosion crack, showing the
transition to ductile fracture.
(b) and (c) illustrate regions
at successively greater distances from the tip (the tip
is to the right in each case);
and show the occurrence of
crystalline deposits, which
form a continuous film
in (c).[9]

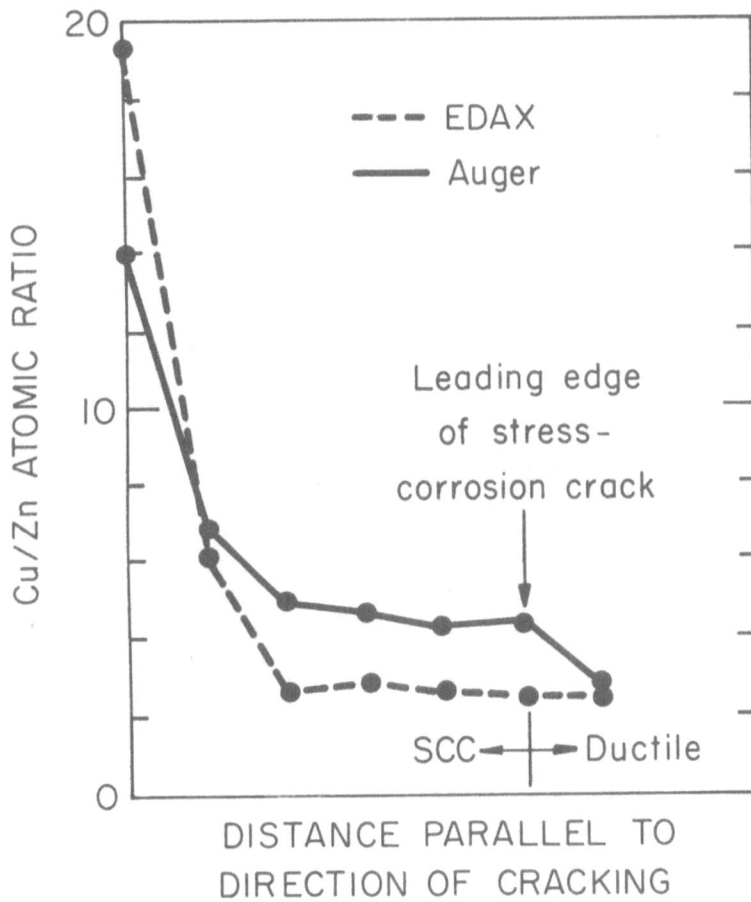

Fig. 9.   Results of EDAX and AES analysis of fracture surfaces
          shown in Fig. 8.[9]

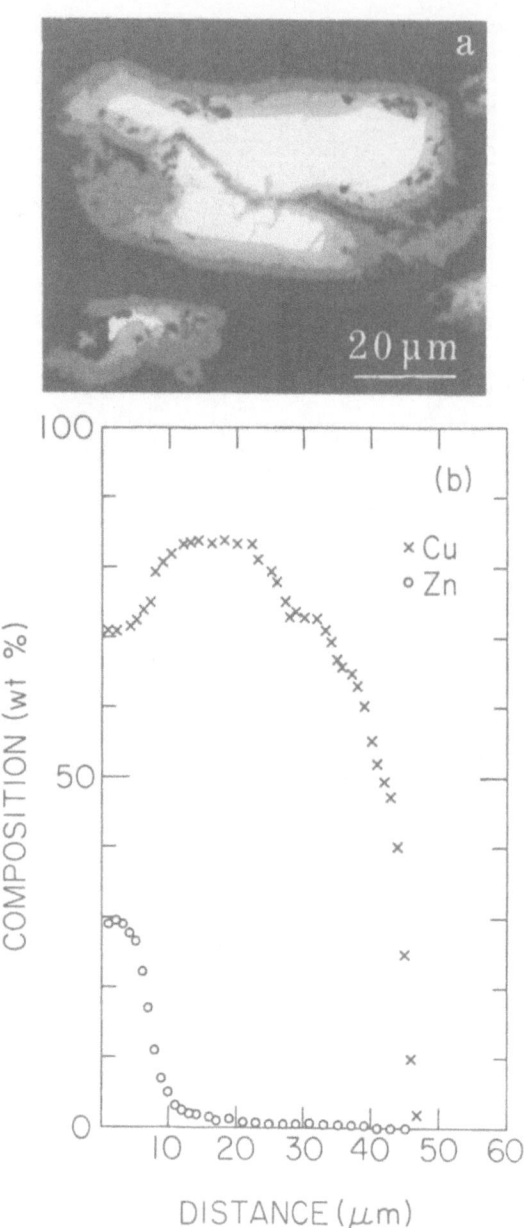

Fig. 10. (a). Optical micrograph of a section of Cu-30Zn filings
which had been immersed for 72 hours in a 15N tarnishing
solution. The tarnish consists of an inner layer of
Cu_2O and an outer layer of CuO. 10(b). Microprobe data
for a sectioned particle such as those shown in (a); the
beam traversed a straight line across the tarnish, beginning in the metal. Note the depletion of Zn in the tarnish
and the plateaux at ~84 and ~ 73% Cu corresponding to Cu_2O
and CuO.[21]

STRESS CORROSION CRACKING IN TITANIUM ALLOYS

Dr. J.C. Scully

Department of Metallurgy, University of Leeds,
Leeds LS2 9JT, England

INTRODUCTION

Stress Corrosion Cracking in titanium alloys at room temperature has been investigated extensively over the last 10 years since it was reported initially in 1966[1]. Phenomenologically it is well described in the Proceedings[2] of the NATO Research Evaluation Conference held at Ericeira. In α alloys containing sufficient aluminium or oxygen fracture in aqueous solutions occurs by cleavage along a plane 14-16° from the basal plane at velocities that can exceed 1 cm/min. In β alloys cleavage occurs on {100} planes up to equally high velocities. In the opinion of the writer the role of hydrogen in these fracture processes was insufficiently emphasized in those Proceedings even though at that time there was sufficient evidence to indicate that hydrogen was of importance. In this paper the role of hydrogen is described in relation to the phenomenological features described in the Proceedings which should be read in conjunction with this.

THE HYPOTHESIS

The original hypothesis, put forward in 1967/68[3,4], proposed that the transgranular cleavage was initiated by hydrogen forming a hydride nucleus which impeded dislocation movement. The conditions for this critical occurrence were discussed at some length. Hydrogen is absorbed by titanium very readily but the existence of a surface oxide film prevents such absorption. Under conditions of stress corrosion cracking in neutral aqueous conditions, it was proposed that hydrogen would be absorbed only during that period of time between the rupture of the passive film at the crack tip and its complete repair, as described in Figure 1[5,6]. The importance of the film repair process was considered to be of extreme importance and the word repassivation was coined to describe it. If repassivation occurred too rapidly then insufficient hydrogen would be absorbed to initiate cleavage. It was supposed that hydrogen absorption would occur rapidly when the film was first fractured and that the rate of absorption would fall off rapidly as the film repaired.

The role of the environment and of the alloy in the repassivation process was discussed in detail[7]. The environment would contain species that promoted repassivation, e.g. OH^- and inhibitive ions, and others that delayed repassivation, e.g. aggressive species indluding halides, some of which could cause hydrogen entry and cracking. The ratio of passivating to de-passivating species was of critical importance in affecting the length of the repassivation time, t_p, which would also be affected by the value of potential and other variables, such as temperature. The alloy would affect the repassivation process through its composition which would have some effect upon the adsorption processes occurring. Most elements would lengthen the repassivation time of titanium if it could be assumed that their effect derived from the electrochemical characteristics of the pure element. Tantalum might shorten this time. It is not always possible and certainly not always sensible to attempt to separate electrochemical aspects of alloying from metallurgical aspects. Alloying elements would alter the slip mode and the creep mode of titanium, both of which were important in determining the occurrence of cracking.

The structure of the alloy was considered to be of prime importance in determining susceptibility. The effect of aluminium additions in promoting co-planar arrays of dislocations was first reported in 1968[4], while the tendency of such additions to raise the average hydride size was reported in 1967[8]. Hydrogen absorption led to hydride formation at the crack tip which raised the local value of the cleavage stress/shear stress ratio[9] in that region because dislocation movement was impeded. A cleavage crack was initiated, a process rendered more easily by the co-planarity of the slip mode The cleavage crack would propagate for mechanical reasons until arrested by a grain boundary or by an occurrence of plastic relaxation. This point is of some importance since it emphasized that the high velocities observed arose from the poor mechanical properties of the particular alloy lattice rather than from a rapid embrittlement process. Crack propagation could go at a higher speed than hydrogen diffusion.

OBJECTIONS TO THE HYDROGEN MECHANISM

The ideas outlined briefly above were not well received in the U.S.A. Objections to the hydrogen hypothesis fell into several categories which are now described.

1.   The observation[e.g. 10] that cathodic polarization could prevent crack initiation and arrest crack propagation in titanium alloy specimens in neutral chloride solutions led to an attempted analogy with high strength steels. The argument put forward was that cathodic polarization should accelerate the onset of cracking if hydrogen was responsible for cracking. The rebuttal of this argument referred[9] back to the situation described in Figure 1. Cathodic polarization would increase at the metal surface the ratio of passivating to de-passivating species (OH^-/Cl^-) and promote more rapid repassivation. The important process of film formation would cause crack arrest. This rebuttal was examined by investigating the role of potential upon crack velocity in 11N HCl, an environment in which no film formation would be expected. In 11N HCl the velocity was constant with respect to potential. Other experiements under open circuit conditions showed[9] that in 3 wt.% Na Cl solutions cracking did not occur in dynamic straining experiments if the pH was increased to 13. This result reinforced the concept that crack propagation would not occur if the OH^-/Cl^- ratio at the metal surface was sufficiently high.

The experiments described above do not show that hydrogen is responsible. What they do show is the importance of film formation in controlling the crack propagation process. With the hypothesis as described, such film formation under cathodic polarization conditions would prevent crack propagation by reducing the rate of hydrogen absorption.

2.   Facture in non-hydrogen containing environments which gives similar cleavage fractography has sometimes been cited as indicating that hydrogen is not responsible for such fractures in hydrogen-containing environments. Stress corrosion in CCl_4 is an example. It has been argued[10] that the residual water content is responsible for cracking in this medium. The

susceptibility of an α-alloy declined as the water content was lowered[10] although experiments were not done below a content of ca. 200 ppm. fractures in fused salt[11] and in liquid metal[12] must also be accounted for. While hydrogen may be responsible in fused salt cracking it seems very unlikely that hydrogen is responsible for liquid metal embrittlement cracking which occurs at extremely high velocities (500 cm/sec). Instead, all that can be said is that the cleavage fractures observed in α alloys do not have to have a single cause. Lowering the cleavage stress/shear stress ratio is the necessary effect to cause a transition from a ductile to a brittle fracture. While the hydrogen hypothesis seeks to explain the transition as arising from absorbed hydrogen there is no reason that other mechanisms might not produce a similar result. A reduction in the cleavage stress by a stress-sorption mechanism in liquid metal[13] could be expected to provide an identical fracture. One important argument for hydrogen, as will be discussed below, is that the fracture observed is mechanical in origin. There is nothing about the fracture that is unique to stress corrosion in aqueous environments. Instead it is observed to occur whenever the mechanical condition at the crack tip is changed so that cleavage fracture occurs in place of a ductile fracture. The occurrence of such a fracture in situations where hydrogen is unlikely to be present (if such an assumption can ever be made) would indicate merely the general nature of the fracture transition. It does not in any way eliminate the role of hydrogen.

3.   The maximum crack velocities observed are much higher than the apparent diffusion rate of hydrogen and this discrepancy has caused difficulty in the acceptance of the hydrogen hypothesis. The maximum velocities observed e.g. >1 cm/min corresponds to $>1.6 \times 10^{-2}$ cm/sec. The diffusivity D is ca. 10^{-10} cm^2/sec. In 1 second a hydrogen atom will travel ca. 10^{-5} cm, corresponding to a velocity of 10^{-5} cm/sec. In 0.000 1 sec the corresponding distance is 10^{-7} cm and velocity 10^{-3} cm/sec. Even if cracking is considered to occur over 10^{-7} cm, about 3 atomic diameters, the velocity calculated is still much less than the measured maximum. This point has, as a general one, been considered by Johnson[14] who discusses it in relation to the ratio D/V, Diffusivity, D, and maximum crack velocity, V. In order to account for a hydrogen mechanism where the kinetics appear to invalidate such a cause, there are 2 possible mitigating circumstances, either or both of which may be applicable. The value of D may be higher than is assumed. At the tip of a propagating crack plastic deformation is occurring and such a process may

increase hydrogen diffusivity. Hydrogen dislocation-sweeping mechanisms have been discussed[15]. Alternatively, if cleavage cracks can run through unembrittled material before being arrested then it is possible to invoke a "long range" mechanism[10] in which a small amount of embrittlement produces a large amount of cleavage. In this case it is not necessary for hydrogen diffusion to occur at speeds comparable to the maximum crack velocity. While it is difficult to prove that such a mechanism occurs, as a general observation it is possible to notice that the highest crack velocities do occur in these alloys with the lowest work-hardening coefficients and lowest ductilities and which might therefore be expected to exhibit the greatest difficulty in arresting propagating cleavage cracks. Hydrogen can be detected in fracture surfaces[16] and appears to occur in local, discrete regions, an observation that would be predictable from a proposed "long range" effect. In Ti-8Al-1Mo-1V alloy Boyd[17] has observed that at a constant velocity the length of cleavage facet observed can vary from 5-20 μm to 50-200 μm, a report that emphasizes the mechanical aspect of this fracture process, since the chemical component in the Stage II cracking region will be constant. The report[18] of identical cleavage in a Ti-Al alloy thermally hydrided and broken in air could also be construed in a similar way: fracture is initiated near to a region that has a hydrogen concentration but propagates into regions that contain less hydrogen.

The kinetic problems presented by the D/V ratio constitute the major obstacle to the hydrogen hypothesis. The two possible ameliorating conditions indicated above can overcome the obstacle but it must be emphasized that a considerable amount of experimental work is required to be done before a clear picture will emerge of this particular feature. Measurements of diffusivity in deforming lattices, quantitative determination of hydrogen, the transport mechanisms of hydrogen and the influence of metallurgical variables are all important and general areas of which detailed information is required. The formation of hydrides during crack propagation has not been observed.

POINTS IN FAVOUR OF THE HYDROGEN MECHANISM

1.   For many years embrittlement experiments have been reported[4,9,10,19] all of which have shown similar results. Exposure of alloys to an aggressive

mixture of CH_3OH and HCl causes intergranular corrosion which is accompanied
by hydrogen absorption. Subsequent fracture of such specimens in air results
in fluted fracture[19] in titanium and fluted fracture with cleavage in
Ti-Al[9] and Ti-0 alloys[20]. A delay between corroding specimens and breaking
them, described as ageing, results in a fracture with decreasing amounts of
cleavage, a tendency accompanied by an increasing elongation to fracture.
After a sufficiently long time no cleavage is seen. These changes are
described in Figure 2. The point that has been emphasized is that the only
fracture arising from dissolution is the intergranular fracture. The cleavage
and fluted fracture are mechanical in origin and do not occur in the embrittlement experiments if sufficient ageing time has elapsed. There is no need
therefore to seek explanations based upon preferential dissolution since the
transgranular fractures are not caused by an anodic process. The ageing
effects can be explained only as a result of effects associated with absorbed
hydrogen which is concentrated immediately after exposure but is dispersed
during the ageing process. These ageing experiments demonstrate very
clearly the role of hydrogen in producing transgranular cleavage. They have
been reinforced by the observation[18] that a thermally hydrided Ti-Al alloy
exhibited identical cleavage when broken in air.

2.   Much work published since the hypothesis was put forward has supported
it. In addition to the hydrided Ti-Al observations referred to above, Boyd[17]
has shown that a hydride phase is readily precipitated by plastic deformation
even though the hydrogen content of the matrix is relatively low (200ppm).
Hayden and Green[21] have shown that cracking occurs much more readily when
specimens are loaded in Mode I than when loaded in Mode III. Under conditions
of Mode I the triaxial hydrostatically stressed region ahead of the crack
tip would be expected to be a region to which absorbed hydrogen would be
drawn whereas no such accumulation under the shearing movements induced by
Mode III loading would be predicted. Gray[22] has shown that, in hot salt
cracking, hydrogen is readily absorbed from the gaseous humid atmosphere
even when the humidity is very low. By using an ion bombardment technique
with a mass spectrometer he obtained profites of absorbed hydrogen in the
fracture surfaces of specimens. A residual level of 100ppm went up to a
value in excess of 10,000ppm. Increasing the humidity of the environment
by several orders of magnitude produced no change, a result that emphasized
the efficient "gettering" quality of the titanium metal surface. Perhaps
this point applies also to room temperature cracking where hydrogen evolution
is observed neither in aqueous nor in methanolic cracking although there is

no question that H^+ ions are being discharged. This discharge process in general produces hydrogen which can be divided into a ratio of hydrogen absorbed/hydrogen evolved. In stress corrosion cracks in Al, Mg and Fe alloys hydrogen evolution is frequently observed. For titanium the non-appearance of hydrogen evolution indicates that the proportion of absorbed hydrogen must be a large part of the total hydrogen that is discharged.

3.   The proposal[23] that cracking occurs by anodic dissolution appears to an untenable proposition. The current density required to dissolve titanium at rates >1 cm/min is >$100A/cm^2$ and such values are observed only in electrochemical machining. In such processes the solution is pumped past the dissolving metal surface at very high velocities in order to avoid concentration polarization. In stress corrosion studies all the evidence suggests that the volume of liquid at the tip is isolated and of different pH from the bulk solution. A pH of 1.7 has been measured[24] for a titanium alloy in 3% NaCl of pH 7. From this evidence it does not seem conceivable that the high solution velocities of electrochemical maching are occurring during stress corrosion propagation. It would be possible only if the hydrolysis rate constants permitted very rapid solution transitions, and there is no evidence for this. Dissolution processes also tend to be described as continuous processes, analogous to an acid saw cutting through metal, whereas the transgranular stress corrosion cracking of titanium alloys occurs as a discontinuous process. This has been recorded[25] by an acoustic emission technique, where the frequency of events corresponds approximately to the grain size. The irregular nature of the propagation is self-evident from the examination of a fracture at either a low magnification or a high magnification. Typical examples are shown in Figures 3 and 4. The cleavage is accompanied by a fluted fracture[26], a low energy tearing process. The largest proportion of the propagation time is the initiation of the cleavage which then occurs rapidly in the usual way and gives place to the fluting process after which cleavage is re-initiated. The conditions leading to cleavage in a stress corrosion process have been discussed fully previously[27].

4.   The slow strain-rate hydrogen embrittlement of α-titanium alloys has been known for over 15 years[28]. It is associated with hydride formation in regions most heavily plastically deformed when lattices contain a supersaturation of dissolved hydrogen. It occurs at relatively low strain-rates which correspond to those which cause stress corrosion cracking in titanium alloys[29].

ADDITIONAL MECHANISTIC POINTS

5.   The importance of film formation is illustrated by details of **Figure 5** as described previously[10]. Where film formation is possible cracking is not observed at low strain-rates. Where film formation is not possible cracking occurs over a wide range of strain-rates. The upper limit in both cases arises from a situation where fracture occurs before crack initiation.

6.   The importance of strain-rate is well demonstrated by Figure 6 where the velocity is plotted as a function of crosshead speed[30]. The velocity is constant over a series of successive 2 mm distances for a given crosshead speed. Double crosshead tests during which the crosshead speed is lowered show that the velocity changes to that corresponding to the lower crosshead speed. Where repassivation is possible arrest occurs at a propagating crack as described in Figure 6.

CONCLUSION

This exposition is intended to redress the apparent imbalance of the Ericeira Proceedings concerning the role of hydrogen in the transgranular stress corrosion cracking of α-titanium alloys. It has been argued that fractographically and phenomenologically the arguments for hydrogen being the principal species responsible is very strong. The outstanding difficulty arises from kinetic considerations and this does not appear to be insuperable. Future work can be expected to resolve this difficulty.

REFERENCES

1. B.F. Brown, Mater. Res. Stds., 66, 129 (1966).

2. The Theory of Stress Corrosion Cracking in Alloys (ed. J.C. Scully), N.A.T.O., Brussels (1971).

3. G. Sanderson, D.T. Powell and J.C. Scully, Fundamental Aspects of Stress Corrosion Cracking (ed. R.W. Staehle, A.J. Forty and D. van Rooyen), p. 638, N.A.C.E. Houston (1969).

4. G. Sanderson, D.T. Powell and J.C. Scully, Corrosion Sci., 8, 473 (1968).

5. J.C. Scully, Corrosion Sci., 7, 197 (1967).

6. J.C. Scully, Corrosion Sci., 15, 207 (1975).

7. J.C. Scully, Corrosion Sci., 8, 771 (1968).

8. G. Sanderson and J.C. Scully, Trans. A.I.M.E. Met. Soc., 239, 1883 (1967).

9. D.T. Powell and J.C. Scully, Corrosion, 24, 151 (1968).

10. J.C. Scully and D.T. Powell, Corrosion Sci., 10, 371 (1970).

11. T.R. Beck, M.J. Blackburn, W.H. Smyrl and M.O. Speidel, Quarterly Progress Report 14, The Boeing Company, Seattle, Washington (1969).

12. M.J. Blackburn, W.H. Smyrl and J.A. Feeney, Stress Corrosion Cracking in High Strength Steels and in Titanium and Aluminium Alloys (ed. B.F. Brown) p. 298, NRL, Washington D.C. (1972).

13. R.M. Latanision and A.R.C. Westwood, ONR Report, N:4162(00), Washington, D.C. (1969).

14. H.H. Johnson, Stress Corrosion Cracking and Hydrogen Embrittlement of Iron Base Alloys (ed. R.D. McCright, J.E. Slater and R.W. Staehle) N.A.C.E., Houston, in press.

15. Hydrogen in Metals (ed. I.M. Bernstein and A.W. Thompson), A.S.M., Ohio (1974).

16. A. Vasel, G. Lapasset, J.P. Laurent, M. Aucouturier and P. Lacombe, L'Hydrogène dans les Metaux, p. 438, Editions Science and Industrie, Paris (1972).

17. J.D. Boyd, Met. Trans., 4, 1037 (1973).

18. D.A. Maumè, E.A. Starke Jr. and R.F. Hochman, Corrosion, 29, 241 (1973).

19. J. Spurrier and J.C. Scully, Corrosion, 28, 453 (1972).

20. T. Adepoju and J.C. Scully, Corrosion Sci., to be published.

21. W.W. Hayden and J.A.S. Green, ref. 15.

22. H. Gray, Corrosion, 25, 337 (1969).

23. T.R. Beck, ref. 2.

24. B.F. Brown, C.T. Fujii and E.P. Dahlberg, J. Electrochem. Soc., 116, 218 (1969).

25. W.W. Gerberich, Fracture 1969 (ed. P.L. Pratt), p. 919, Chapman and Hall, London (1969).

26. B. Cox, Corrosion, 28, 207 (1972).

27. J.C. Scully, Third International Congress on Fracture, PL-IV-222, Dusseldorf (1973).

28. R.I. Jaffee and D.N. Williams, Trans. A.S.M., 51, 820 (1959).

29. D.T. Powell and J.C. Scully, ref. 25, p. 407.

30. T. Adepoju and J.C. Scully, Corrosion Sci., 15, 415 (1975).

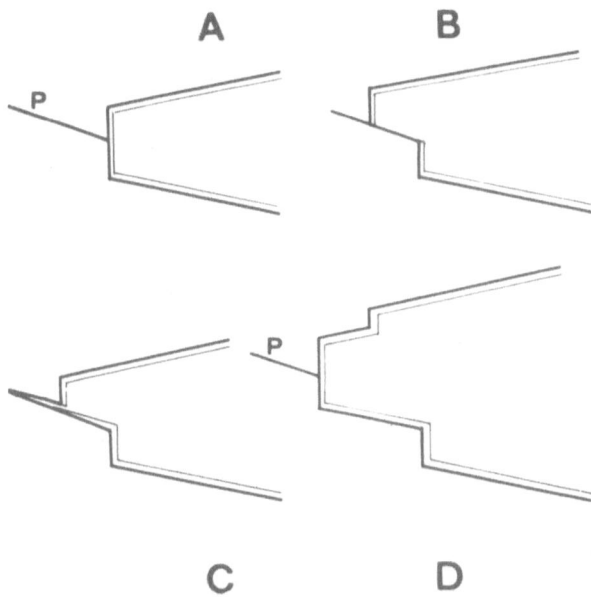

Figure 1   Schematic diagram of the sequence of events occurring at the
tip of a propagating stress corrosion crack.

(a) The tip of a crack. The surface is covered by a protective
film. P is a slip plane.

(b) The slip plane undergoes a shear and new, unfilmed, reactive
surface is formed.

(c) Corrosion attack occurs accompanied by hydrogen absorption.

(d) Repassivation occurs and crack arrest is observed. Embrittlemer
by absorbed hydrogen e.g. the initiation of cleavage must occur
before the film will be re-broken.

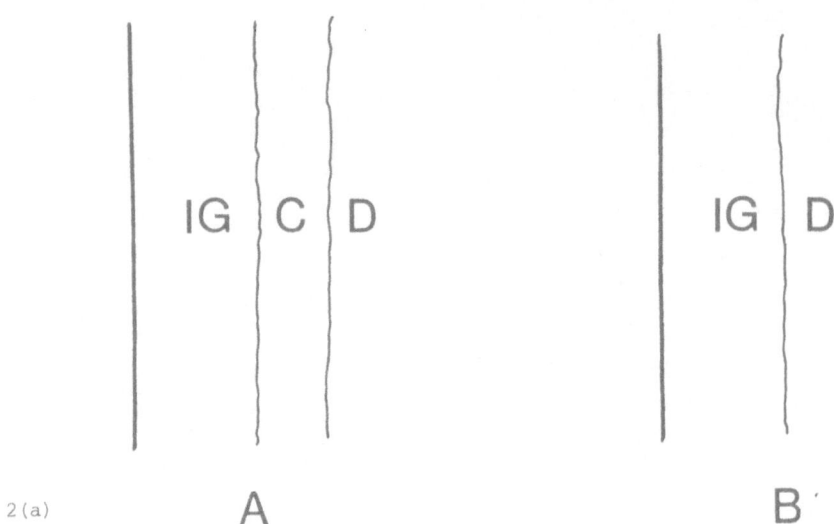

2(a)   A   B

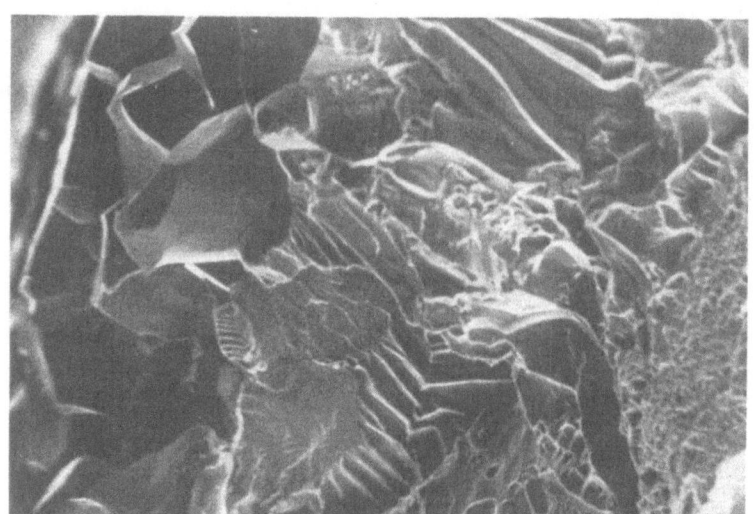

2(b)

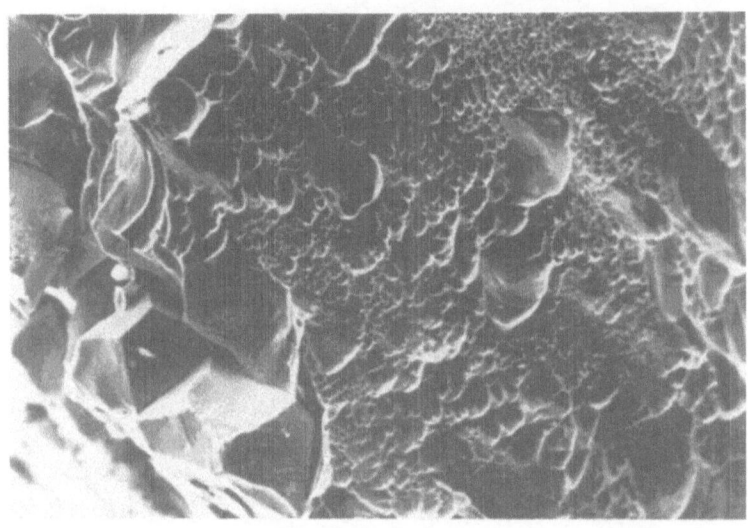

Figure 2   (a) A schematic diagram of the fracture surface of Ti-0 alloy
specimens broken (a) immediately after being corroded in the
unstressed condition in a MeOH/HCl mixture, when an intergranular
fracture caused by dissolution and a dimple fracture are separated
by a narrow zone of cleavage and fluting, and (b) after an ageing
time. Only intergranular and dimple fracture regions are observed.
Fractures corresponding to these situations are shown in
Figures 2(b) and 2(c).

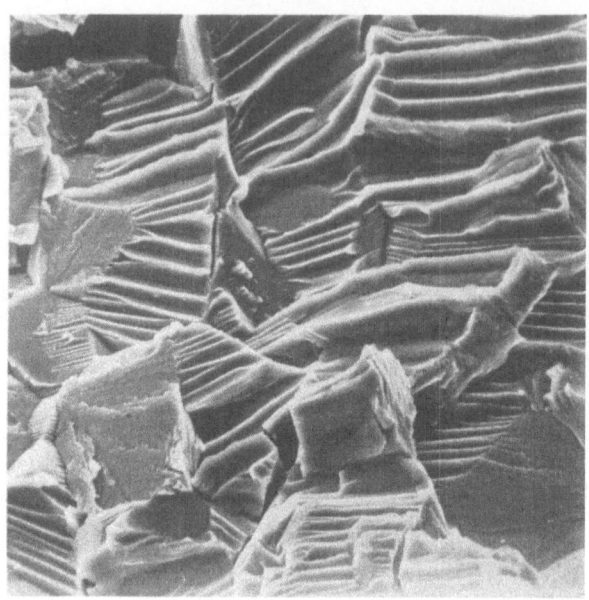

Figure 3 A scanning electron microscopy picture of transgranular stress corrosion fracture in a Ti–0 alloy. A large amount of fluted fracture is visible with some areas of cleavage. 400 x

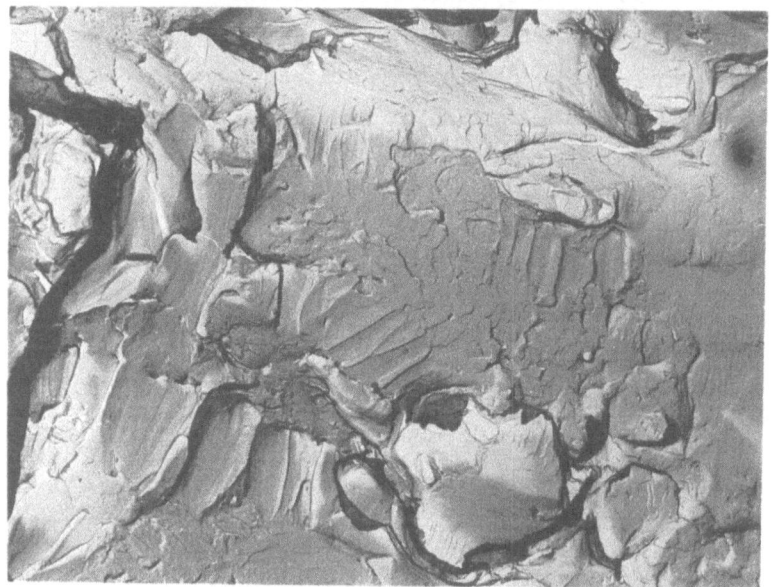

Figure 4 A transmission electron microscopy picture of stress corrosion fracture in a Ti–6Al–4V alloy obtained by a replica technique. The fluted and cleaved parts of the fracture are both visible. 600 x

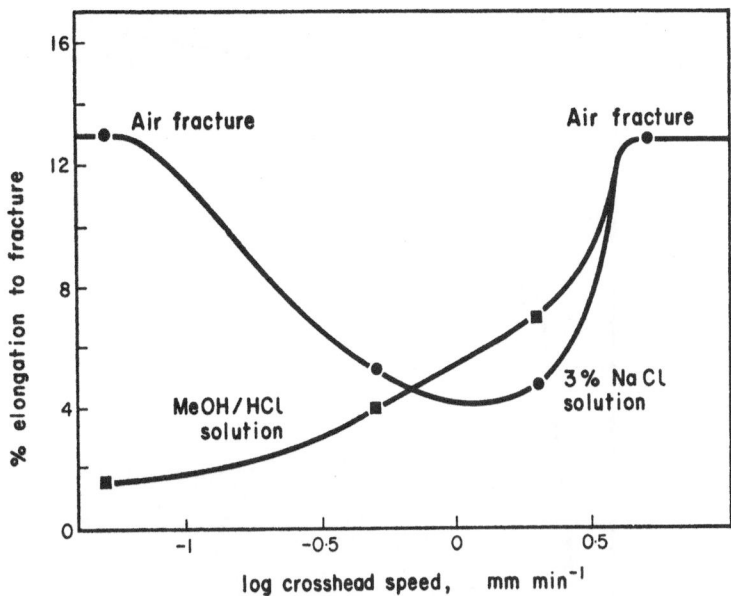

Figure 5   The elongation to fracture of tensile specimens of Ti-5Al-2.5Sn
            alloy in 3% aqueous NaCl and a MeOH/HCl mixture as a function of
            Instron crosshead speed. In the aqueous solution repassivation
            occurs at lower crosshead speeds[10].

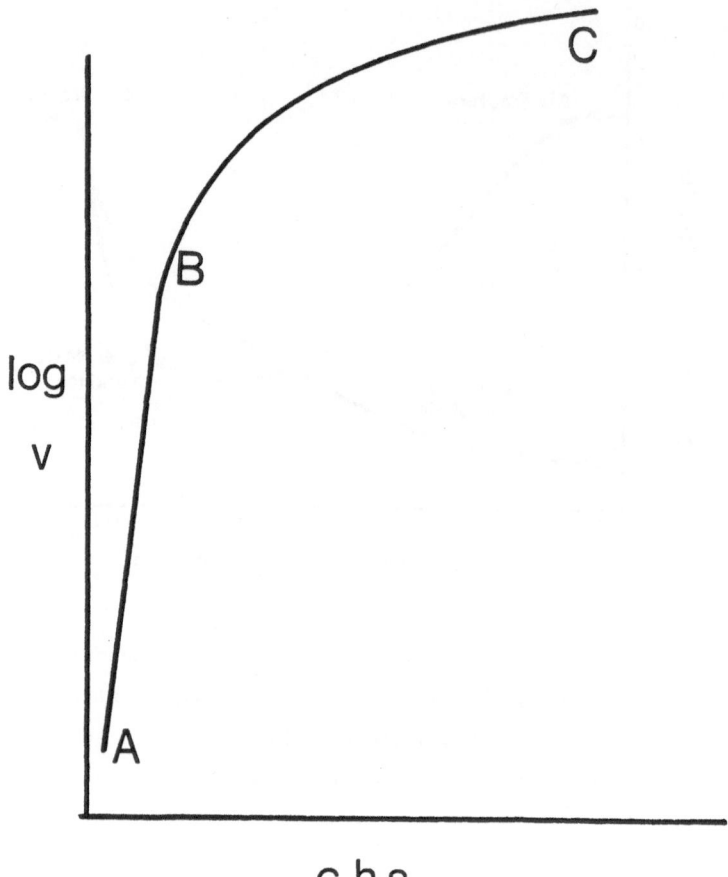

Figure 6   The relationship between measured stress corrosion crack velocity
and Instron crosshead speed[30]. Specimens strained at a constant
crosshead speed exhibited a constant crack velocity. If the
crosshead speed was altered just after crack initiation then the
velocity fell or rose to a value corresponding to the second
crosshead speed. In aqueous 3% NaCl solution repassivation and
crack arrest occurred if the second crosshead speed fell below a
minimum value corresponding to B. Thus A,B,C represents
schematically the values for MeOH/HCl and B,C represents the
values for 3% NaCl

ADSORPTION-SENSITIVE FRACTURE PROCESSES

A. R. C. Westwood
Martin Marietta Laboratories
Baltimore, Maryland   21227, U.S.A.

ABSTRACT

This paper is intended to provide an introduction to the field of adsorption-sensitive fracture behavior. The occurrence, characteristics, mechanisms, and control of such phenomena as liquid metal embrittlement, complex-ion embrittlement, and chemomechanical effects in ceramics and glasses are described and discussed.

1. INTRODUCTION

The presence of adsorbed, surface-active species at regions of stress concentration in solids can markedly influence their fracture behavior. The best known examples of such effects are, of course, associated with the liquid metal embrittlement of solid metals,[1-4] a phenomenon discovered some sixty years ago[1]. But such effects occur in other classes of solids also, for example in silver chloride[5], magnesium oxide[6, 7], soda-lime glass[8], polymethylmethacrylate [9], and naphthalene [10], although the mechanisms involved are likely to be somewhat different for each class of solid.

Nowadays, materials scientists and design engineers are becoming increasingly aware of the existence of such phenomena, because they usually exert a detrimental influence on the lifetime and reliability of components in service. Such recognition is evidenced by the increasing use in design studies of fracture mechanics data derived from tests involving pre-cracked specimens exposed to some relevant operating environment, e.g., an aqueous chloride solution. A start has also been made towards the control of such effects, primarily with a view to minimizing them, but alternatively with the object of utilizing them to facilitate fracture, and hence the chip form processes, in such operations as drilling, shaping, and comminution[11]

This paper is intended to provide an introduction to the field of adsorption-sensitive fracture behavior. Some examples from recent studies will be described, and the current status of mechanistic understanding discussed. The text is based on earlier review articles prepared by the author and his colleagues, and cited as Refs. 3, 11 and 12.

2. LIQUID METAL EMBRITTLEMENT

A. Occurrence and Mechanisms

When an oxide-free solid metal is coated with a liquid metal and immediately deformed in tension, its yield and flow behavior are not significantly affected. Its fracture behavior, however, can be markedly different from that observed in air. In some instances a reduction in fracture stress or strain results, the magnitude of which depends on various chemical and mechanical parameters of the solid metal-liquid metal system.

Embrittlement by liquid metals can occur also in the absence of stress, by corrosion or diffusion-controlled intergranular penetration processes. The most dramatic and catastrophic examples of liquid metal embrittlement are, however, those which appear to be dependent upon adsorption rather than absorption or dissolution effects.

Liquid metal embrittlement (LME) phenomena are potentially of concern whenever liquid metals are used in contact with oxide-free solid metals[4], e.g., during brazing, galvanizing, or heat exchanging operations. Less obvious situations can also lead to unexpected failures, e.g., the fracture of a mercury - containing thermometer in an aluminum boat has lead to cracks in the hull; titanium alloy jet compressor discs in contact with cadmium plated bolts have failed at temperatures above the melting temperature of cadmium[4]; certain leaded steels have exhibited marked reductions in ductility at elevated temperatures, etc.[13].

Fig. 1.   Cleavage failure of a notched, 12 mm diam. aluminum monocrystal stressed in liquid gallium. The fracture surface is predominantly $\{100\}$ [15].

Current thinking is that liquid metal embrittlement results from some chemisorption-induced reduction in the cohesive strength of atomic bonds at regions of stress concentration in the solid metal[2]. Thus, the prerequisites for its occurrence are (i) a tensile stress, (ii) either a pre-existing crack, or some measure of plastic deformation and the presence of a stable obstacle to dislocation motion in the lattice, e.g., a grain boundary or precipitate particle, and (iii) adsorption of the active embrittling species specifically at this obstacle and subsequently at the tip of any propagating crack[2, 3].

However, the factors that determine which liquid metal will actually embrittle which solid metal remain unclear. In general, it appears that to constitute an embrittlement couple, both the solid metal and the liquid metal should exhibit limited mutual solubility and little tendency to form stable, high-melting point intermetallic compounds. Dissolution processes are not thought to be relevant because a suitably prestressed metal specimen will fail virtually instantly on being wetted with an appropriate liquid metal, and presaturating the liquid metal with the solid metal does not influence embrittlement behavior significantly[14].

Embrittlement can occur in all degrees, and the catastrophic embrittlement of an otherwise ductile solid, e.g., the failure by cleavage of an aluminum monocrystal, Fig. 1[15], is only an extreme case. Since even the most active liquid metals affect only the stress and strain at fracture, and not the yield stress, the term embrittlement is used simply to denote a reduction in strength or ductility. It is not intended to imply that the fracture process necessarily occurs in a completely brittle manner.

Since monocrystals can be embrittled, grain boundaries are not essential to the phenomenon. However, liquid metal embrittlement (LME) usually is more severe in polycrystals than in monocrystals. Grain boundaries, of course, serve as stable obstacles to dislocation motion so, assuming a dislocation mechanism for crack initiation, this observation is not unexpected. Failure usually occurs in an intercrystalline manner, presumably because less energy is required than for transcrystalline cleavage in most ductile metals. For highly anisotropic metals such as zinc, however, failure of polycrystalline specimens occurs predominantly by cleavage on basal planes.

To appreciate the general character of LME, it is useful to consider the nature of the interaction between an adsorbed species and a metallic surface[2]. Because the high concentration of mobile conduction electrons within the metal screens out the effects of any adsorbed specie within a distance of a few atomic diameters of the surface[16], it follows that adsorbates should not influence significantly the bulk mechanical properties of oxide free metals. This conclusion is in accord with the general observation mentioned above that the yield or flow behavior of a metal crystal is not significantly affected when exposed even to highly surface active liquid metals. Adsorbates may influence fracture behavior, on the other hand, because the propagation of a surface-initiated crack essentially involves the consecutive rupture of surface bonds.

Consider, then, the possible effects of an adsorbed surface active liquid metal atom B at a crack tip in a stressed solid metal A, Fig. 2. Chemisorption of atom B will cause some variation in the tensile strength, σ, of the bond between atoms A and A_o constituting the crack tip. If it is an embrittling atom, σ will be reduced. Because of conduction electron screening, however, it is unlikely that atom B will be able to influence the strength of the bonds across any slip plane for a sufficient distance from the crack tip to affect the ease of dislocation motion away from the crack. For this reason, the shear stress, τ, required to move dislocations on slip planes in the vicinity of the crack tip should not be significantly altered by the presence of atom B. Adsorption of an embrittling

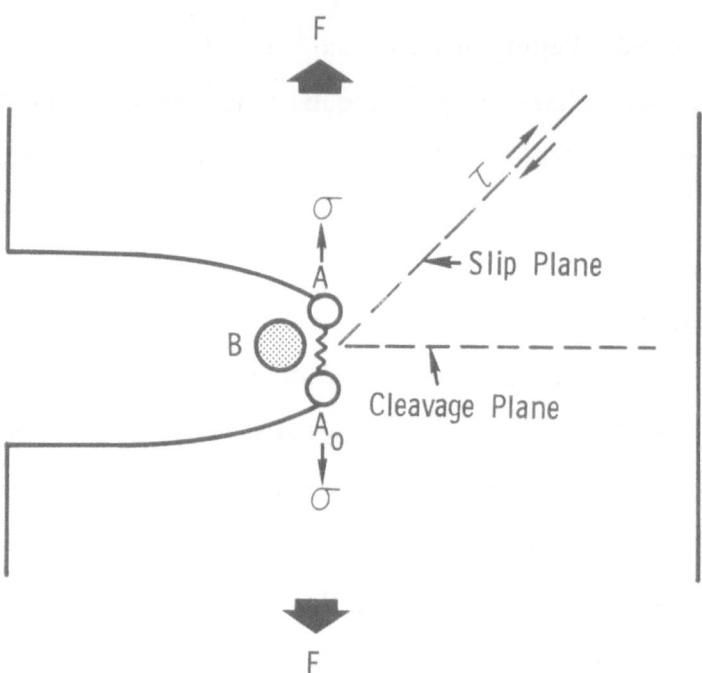

Fig. 2   Schematic of a crack in a solid subjected to an increasing
force F. The bond A-A_0 constitutes the crack tip. B is
a surface active liquid metal atom[15].

species leads, therefore, to a decrease in the ratio (σ / τ). On the basis of

arguments presented by Kelly et al.[17] this decrease will then be manifested

as an increased tendency for the crack to propagate by cleavage rather than

ductile shear, and the severity of embrittlement observed will depend upon

the magnitude of the reduction of (σ / τ). An extremely active liquid

metal can reduce (σ / τ) sufficiently to cause failure predominantly by

cleavage even in normally ductile f.c.c. metal such as pure aluminum, see Fi

    Conversely, of course, it should be possible to increase the ratio

(σ / τ) by adsorbing at the crack tip some liquid-metal species which interacts

at the A-A_0 bond so as to increase σ. An element known to form high

melting point (strongly bonded) intermetallic compounds with the solid metal

might be expected to act in this manner, and the possibility of utilizing this

feature to inhibit LME will be discussed below.

B.   Influence of Physical and Chemical Variables

The severity of embrittlement by mercury of a series of copper base alloys increases as the stacking fault energy or ease of cross slip decreases, Fig. 3(a). It might be concluded, therefore, that alloying additions control embrittlement principally by affecting the parameter τ, and that the chemical nature of the solute has no particular effect. However, the electron/atom ratio both determines the stacking fault energy and also affects the elastic constants of an alloy. Thus, since alloying can also affect σ, it may be this factor which actually controls susceptibility in some instances. Significantly, the data shown in Fig. 3(a) exhibit an equally good correspondence with the electron/atom ratio, Fig. 3(b)[2].

The susceptibility of polycrystalline zinc to embrittlement by liquid mercury is also markedly influenced by alloying. For example, additions of as little as 0.2 a/o copper or gold in solid solution significantly reduce its fracture stress, Fig. 4(a)[19]. It was also found that the critical resolved shear stress, τ_C, for chemically polished zinc monocrystals stressed in air increased with copper and gold content, as shown in Fig. 4(b). It appears that in this system, therefore, the increased susceptibility to embrittlement is not related to solute-induced changes in stacking fault energy or bond strength -- because of low concentration of solute present -- but more directly to the variation in τ_C. Increasing τ_C (decreasing σ/τ) inhibits the relaxation by plastic flow of stress concentrations at grain boundaries, and so facilitates crack initiation in the presence of mercury.

One important factor which requires much more attention is the variation in susceptibility to embrittlement with chemical composition of the liquid metal. A useful approach to this problem involves dissolving the potentially active metal in an "inert-carrier" liquid metal of lower melting point. This approach has allowed both evaluation of the embrittlement behavior of a number of new systems, and comparisons to be made of the

232

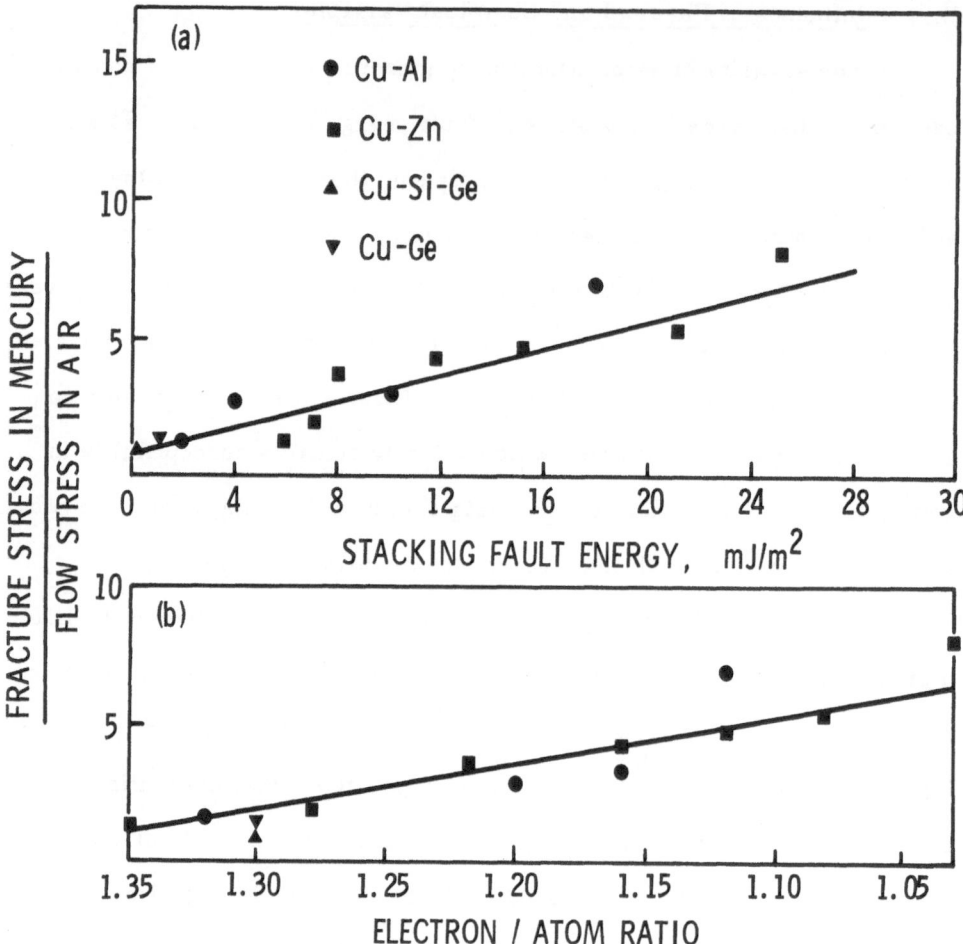

Fig. 3.   Embrittlement of copper base alloys as a function of (a) stacking fault energy[18], and (b) electron/atom ratio[2].

influence of several elements effectively in the liquid state at the same temperature[20]. The need for such a technique becomes evident when it is appreciated that the direct investigation of many potentially interesting solid metal-pure liquid metal couples is not feasible because, at the melting temperature of the pure liquid metal, the solid metal is either too ductile to maintain the stress concentrations necessary to initiate and propagate a brittle crack, or is excessively soluble in the liquid metal, resulting in crack blunting.

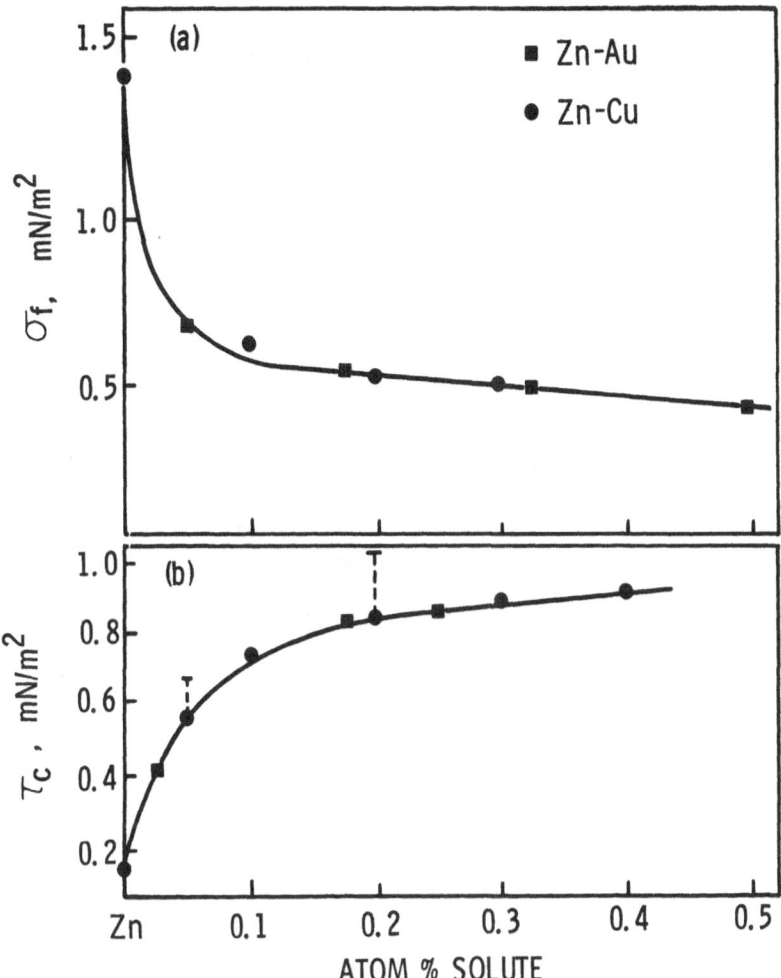

Fig. 4.   (a) Effect of solute content on fracture stress, σ_F, of
polycrystalline zinc in liquid mercury. The grain diameter
of the zinc was ~1 mm. (b) Variation of critical resolved
shear stress, τ_C, with solute content for zinc monocrystals [19].

Recent studies with liquid metal solutions have indicated that LME

is not nearly as specific as had been previously thought. For example,

polycrystalline pure aluminum is only slightly embrittled by mercury at room

temperature, but additions of as little as 1-3 a/o of a number of elements to

the mercury produce marked effects on the severity of embrittlement[20],

Fig. 5.

The severity of embrittlement by liquid metal solutions can also be extremely temperature sensitive[2, 21]. Figure 6 illustrates, for example, the occurrence of brittle-to-ductile transitions in polycrystalline pure aluminum, the critical temperatures, T_c, for which are determined by the composition of the mercury-gallium solution environment. Similar effects have been observed also for silver and brass[21]. It appears that previous analyses of apparently similar transitions occurring in bcc or

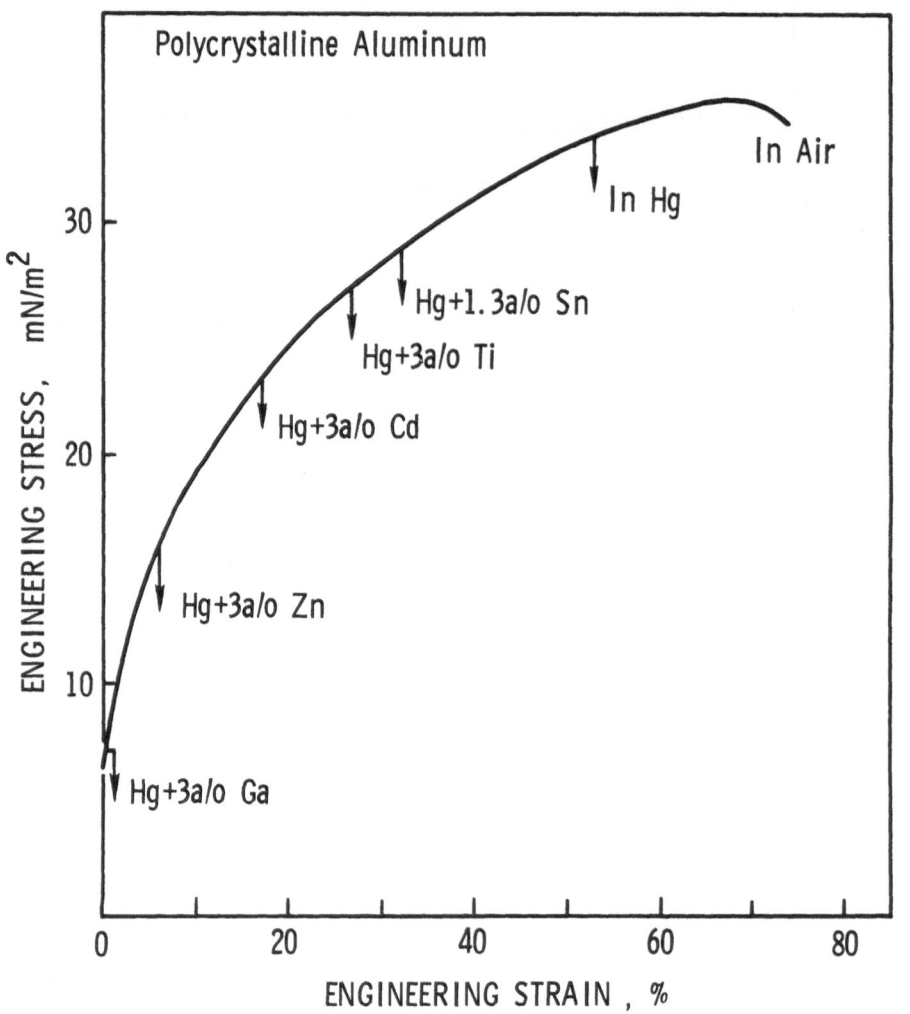

Fig. 5.   Embrittlement of polycrystalline pure aluminum by various mercury solutions [20].

hcp metals when tested in inert environments[22] are inappropriate for such environment-sensitive transitions because the assumptions on which they are based are not valid in the latter case. The classical assumptions are (i) that the temperature dependence of the yield stress is the controlling factor, and (ii) that T_c is the temperature at which the yield stress equals the fracture stress. For fcc metals such as aluminum, however, the yield stress decreases only slightly with temperature over the temperature range involved, and a significant amount of plastic deformation occurs before embrittlement results, even below T_c. Thus, the yield stress does not

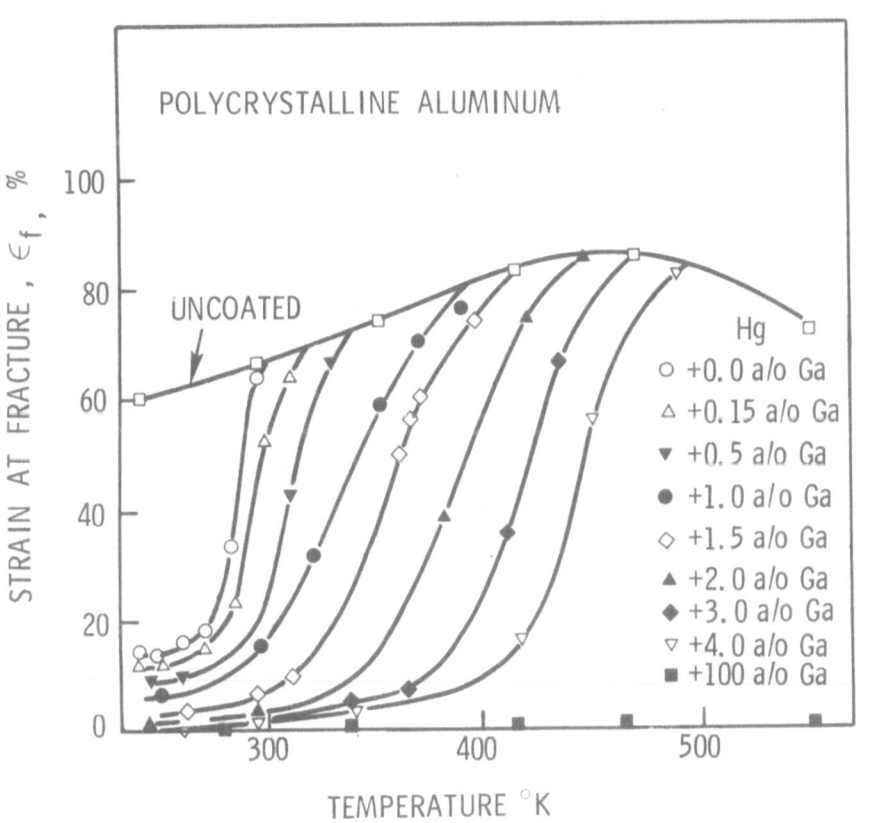

Fig. 6.   Temperature dependence of strain at fracture for polycrystalline aluminum speciments in mercury, gallium and mercury-
gallium solutions [21].

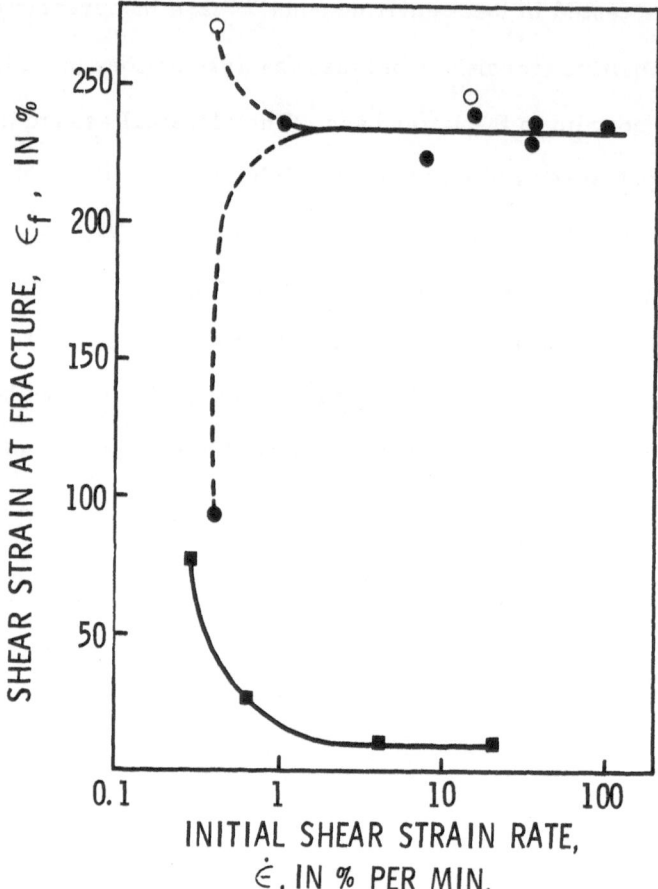

Fig. 7.   Strain rate dependence of the shear strain at fracture for (i)
lower curve, 1mm dia. zinc monocrystals coated with gallium[23]. Note "plasticizing" effect at low strain rates.
(ii) upper curves, 6mm square zinc monocrystals partially
coated with gallium (●) or uncoated (O)[24].

equal the fracture stress at any temperature of relevance to this work. A

new analysis has been developed[21], therefore, which assumes that such

transitions are associated with the temperature-sensitive behavior of the

ratio (σ_e/τ), σ_e being the effective tensile fracture stress of atomic bonds

in the presence of an adsorbed and embrittling liquid metal species. T_c is

the temperature at which (σ_e/τ) increases above some critical value.

While τ is a relatively temperature-insensitive parameter, σ_e is tem-

perature sensitive because adsorption (desorption) is a dynamic and thermally
activated process.

The strain rate sensitivity of LME is not yet reliably established
or understood. Shchukin et al.[23] found that the strain at fracture, ϵ_F,
for 1 mm diam. zinc monocrystals oriented for single slip and coated
with liquid mercury, gallium or tin was markedly dependent upon strain
rate, $\dot{\epsilon}$. For $\dot{\epsilon} \sim$ 10-15% per min., all three liquid metals reduced
ϵ_F from that in air, but at very low strain rates (10^{-1} - 10^{-3}% per min.)
the ductility of wetted crystals increased sharply. More recent work[24]
was unable to reproduce these results, however. Indeed, it was found that
neither gallium nor mercury produced any significant embrittlement of
6 mm square monocrystals oriented for single slip and carefully handled to
prevent accidental damage. Moreover, not only was the plasticizing effect
observed by Shchukin et al.[23] absent, an opposite effect was found, Fig. 7.
For uncoated crystals at low strain rates, ϵ_F did increase, this effect
being interpreted in terms of the simultaneous deformation and recovery
of zinc at room temperature. But for coated crystals, ϵ_F decreased. The
discrepancy between the observations of Shchukin et al., who used 1 mm
diam. crystals, and later workers probably arises from the fact that it is
difficult to handle 1 mm diam. zinc monocrystals without accidentally
damaging them. They then deform in an inhomogeneous manner, forming
kink bands which serve as crack initiating barriers to slip in the presence
of an embrittling liquid metal. Consequently, such specimens fail pre-
maturely. However, when slow strain rates are used at room temperature,
some of this damage can anneal out -- hence the apparent "plasticizing"
effect. Larger diameter crystals, on the other hand, are essentially damage
free, and so are not embrittled when tested at normal strain rates.

When tested at very slow rates, however, kink bands are formed by dislocation climb. These provide dislocation barriers and lead to embrittlement.

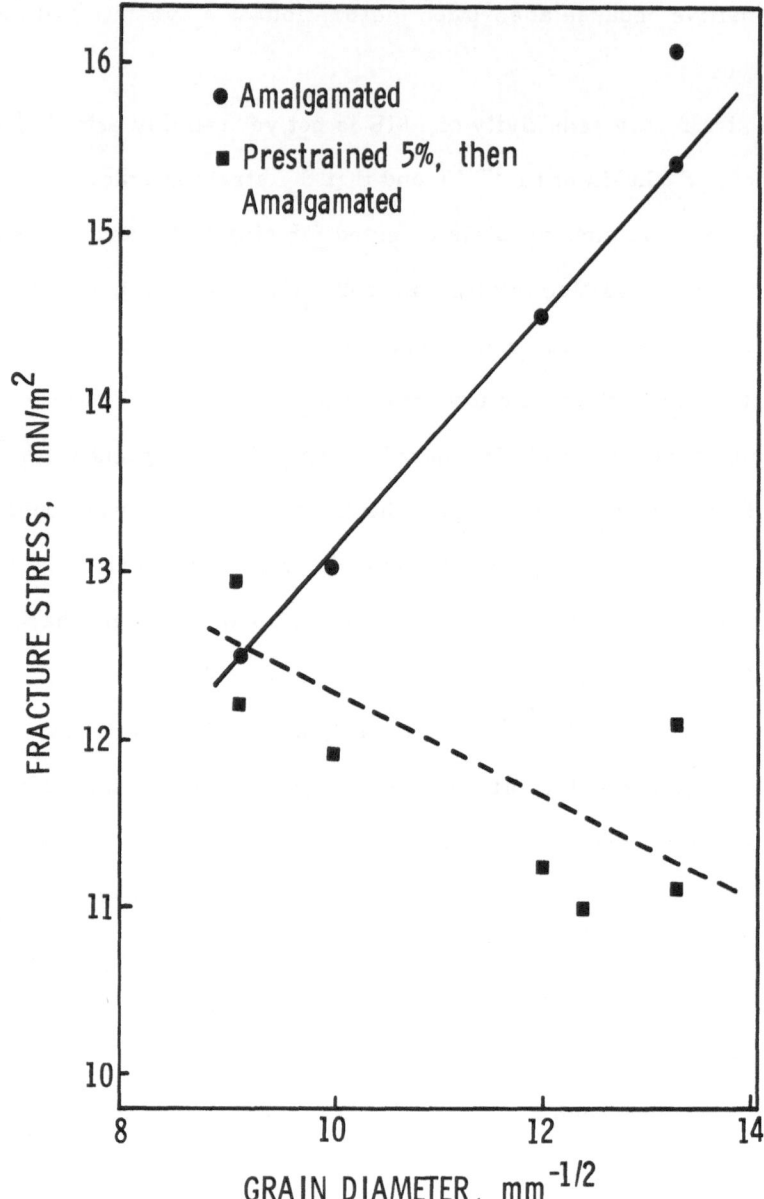

Fig. 8. Effects of grain size and prestrain on fracture stress of amalgamated Al 5083 alloy specimens. Note that the data from prestrained specimens do not conform to a Petch relationship [26].

For polycrystalline metals, increased strain rates might be expected to produce greater susceptibility to LME, because increasing $\dot{\epsilon}$ raises τ , and this decreases the ratio (σ / τ). For a notch-insensitive solid metal, however, an upper limit for embrittlement will occur when the liquid metal can no longer keep up with the propagating crack.

Prestraining also affects susceptibility to embrittlement, normally increasing the fracture stress and decreasing the strain at fracture [25]. However, the effects of prestrain are not always those expected, as indicated in Fig. 8, which presents some data obtained with a strain-aging aluminum alloy[26]. The data for the prestrained and amalgamated aluminum alloy specimens clearly do not conform to a Petch-type relationship, since the slope of the data indicates a negative fracture energy ! Other workers have found that severely cold-worked (> 50%) specimens of α-brass, iron and Fe-Si alloys exhibit little susceptibility to LME, and fail transgranularly[27] For deformations of up to about 25%, however, increases in embrittlement were observed, and failure was intergranular. It has been suggested that fragmentation of the original grain boundaries by extensive cold working is responsible for the observed reduction in susceptbility and change in fracture mode.

C. Control of LME

Given that to constitute an embrittlement couple both the solid metal and the liquid metal should exhibit limited mutual solubility and little tendency to form stable, high melting point compounds [28, 29], two possible ways of inhibiting LME may be suggested:[11]

(i) add some soluble element B^* to the liquid metal B which embrittles solid metal A; B^* being chosen because of its known tendency to form stable compounds with A. Inhibition may then result either from preferential adsorption of B^* on A, screening A from B, or from the formation of relatively strong A-B*-A bonds at crack tips which counter the weakening influence of the element B on A-A bonds.

Several examples of the usefulness of this approach may be cited. For instance, barium is slightly soluble in liquid mercury at room temperature and is known to form stable intermetallic compounds with zinc, such as $Zn_{13}Ba$. Thus, the addition of a small amount of barium to mercury might be expected to reduce the severe embrittlement of zinc by mercury, and such an effect has been observed[30]. The fracture stress of poly-zinc specimens (~1 mm grain diam.) amalgamated with pure mercury was determined to be ~6 ± 1.5 MN/m^2, whereas that of similar specimens coated with mercury containing 0.4 a/o barium was ~ 10 ± 3.2 MN/m^2 -- about a 70% i nprovement. Similarly, the embrittlement of polycrystalline copper by bismuth is well known. But, according to recent work by Yuschenko[31], additions to bismuth of tin (which forms intermetallic compounds with copper) markedly reduces the severity of embrittlement. Other examples include the addition of indium to reduce the mercury embrittlement of silver and α-brass[21].

(ii) alloy some soluble element A* with the solid metal A; A* being chosen because of its known tendencies to form strong intermetallic bonds with the embrittling liquid metal element B. A possible example of such behavior is the addition of ~30 a/o of gold to silver, which effectively inhibits the embrittlement of silver by gallium, but not by mercury[32]. Gold forms high melting point intermetallic compounds with gallium, but not with mercury.

Another way of controlling the severity of LME is by changing the operating temperature for, as noted above and evidenced in Fig. 6, LME is extremely temperature sensitive.

For more complete reviews of the characteristics and mechanisms of LME, see Refs. 2, 4, 28, 29 and 33.

3. COMPLEX-ION EMBRITTLEMENT

In contrast to metals, and because of the much lower concentration and mobility of charge carriers present, the electrical influence of a species chemisorbed at the surface of a nonmetal can be felt as much as several microns in from the surface[16, 34]. Potentially, therefore, such adsorbates can affect both near-surface flow and flow-dependent fracture behavior. One intriguing consequence of this is the phenomenon of complex-ion embrittlement.

A. OCCURRENCE AND CHARACTERISTICS OF COMPLEX-ION EMBRITTLEMENT OF AgCl

When polycrystalline AgCl is exposed at room temperature to aqueous environments containing highly charged complex ions, such as 6N sodium chloride presaturated with AgCl, in which the predominant complex species is $AgCl_4^{3-}$, its fracture mode changes from ductile and transcrystalline to brittle and intercrystalline[35, 36]. Both positively and negatively charged complexes can cause brittle behavior, and it has been found that the degree of embrittlement increases with concentration and charge of the critical complex species present in the environment, and is a function of the distribution of charge on the complex.

Metallographic studies on polycrystalline AgCl specimens stressed in complex-containing environments reveal that cracking is initiated where slip bands are arrested at grain boundaries of large misorientation. In Fig. 9(a), for example, intercrystalline cracks (arrowed), which appear dark in transmitted light, have been initiated at each of the blocked slip bands A, B, and C. Once initiated, such cracks propagate along grain boundaries in a discontinuous manner, as evidenced by the "striae" on the intercrystalline fracture surface shown in Fig. 9(b). Such striae are considered to mark arrest positions of the propagating crack.

Tests on monocrystals have established that unnotched specimens are essentially immune to cracking on complex-containing solutions, but that

notched monocrystals are severely embrittled, failing by discontinuous cleavge
It may be concluded, therefore, that embrittlement is not a consequence of
some inherent property of the grain boundary. Nevertheless, boundaries
play an important role in the embrittlement of polycrystalline AgCl by acting
as barriers to glide dislocations, introducing stress concentrations, and
facilitating crack initiation.

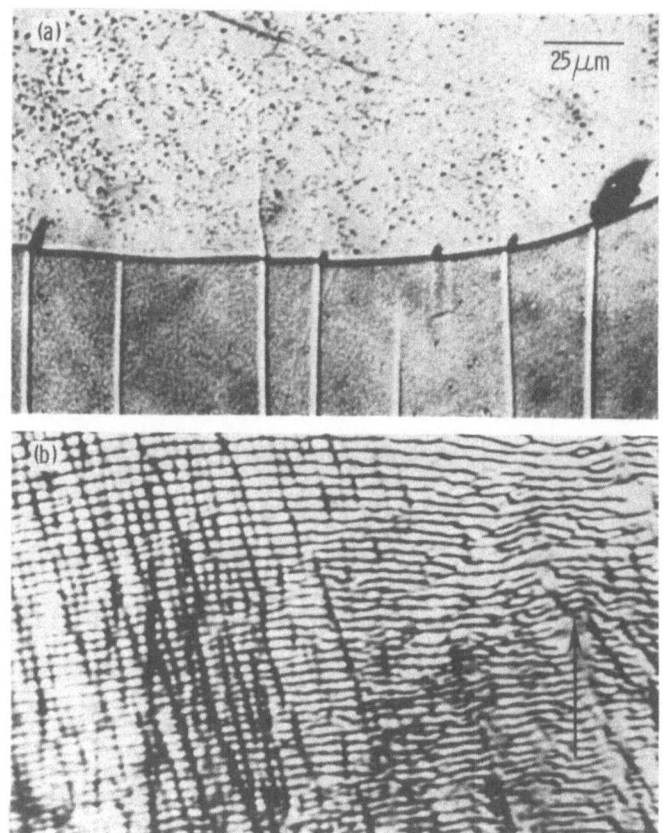

Fig. 9. (a) Polycrystalline AgCl strained in aqueous 6N NaCl containing AgCl$_4^{3-}$ complexes. Cracks are initiated where slip
bands are arrested at the boundary, e.g., at A, B, and C.
Transmitted light [36]. (b) Intercrystalline fracture surface
reveals that crack propagation was discontinuous, apparently
alternating between intercrystalline "cleavage" and plastic
relaxation in the vicinity of the crack tip. The direction of
crack propagation is denoted by the arrow.

B.   Possible Mechanisms of Embrittlement

Present thinking[37] is that the complex-ion embrittlement of AgCl is associated with the repeated formation and rupture of point defect-hardened charge double layers, as illustrated schematically in Fig. 10. The charge carriers in AgCl at room temperature are Frenkel defects, silver interstitials Ag_i^+, and silver vacancies Ag_v^-, both of which are highly mobile at room temperature. Following the arguments of Grimley and Mott[38] and others, therefore, it might be envisaged that the adsorption of complex ions of high negative charge at the surface of an AgCl crystal would result in the development of a compensating positive charge in the immediate vicinity of the surface. This charge, termed the "surface charge", would be produced by the presence of a sufficient concentration of Ag_i^+ ions. A potential balancing negative space-charge, in the form of a more diffuse aggregation of Ag_v^- defects, would then form further into the crystal. This would represent the "near-surface charge", as defined by Lifschitz and Geguzin[39]. Such a situation is shown schematically in Fig. 10(a). Adsorption of positively charged complex ions would be expected to produce a double layer of opposite sense.

Now, the presence of a large concentration of point defects in the surface layers of a specimen would be expected to cause significant surface hardening. Thus, if a specimen exhibiting such a hardened layer were stressed, it is likely that stress concentrations, such as those induced by a pile-up of dislocations at a grain boundary, would cause it to crack (Figs. 10(b) and (c)). Once formed, the crack formed would propagate readily through the embrittled surface layer, but probably would blunt out when it entered the softer space-charge region of the double layer, or the more ductile matrix material. Meanwhile, however, the liquid environment would have entered the crack, the active species adsorbing on its walls and at its tip, and inducing thereby the formation of a new, defect-hardened region around and ahead of the crack. Then, when the region immediately ahead of the crack became sufficiently brittle, the crack would again propagate under the combined action of the applied stress and stresses set up by continued yielding in the plastic matrix, and again blunt out as it emerged from the severely hardened layer. This sequence, illustrated schematically in Figs. 10(c) through (g), could be repeated until the specimen failed.

Much experimental evidence in support of this mechanism for complex-ion embrittlement is now available[37]. For example, the discontinuous failure process described would produce striated fracture surfaces, and evidence for this was shown in Fig. 9(b). In addition, the postulated surface hardening associated with the adsorption of complex ions has been revealed by microhardness studies. Finally, according to the model, the rate of crack propagation, and hence the time to failure t_F, should be controlled by the rate of formation of the brittle layer ahead of the crack. Under

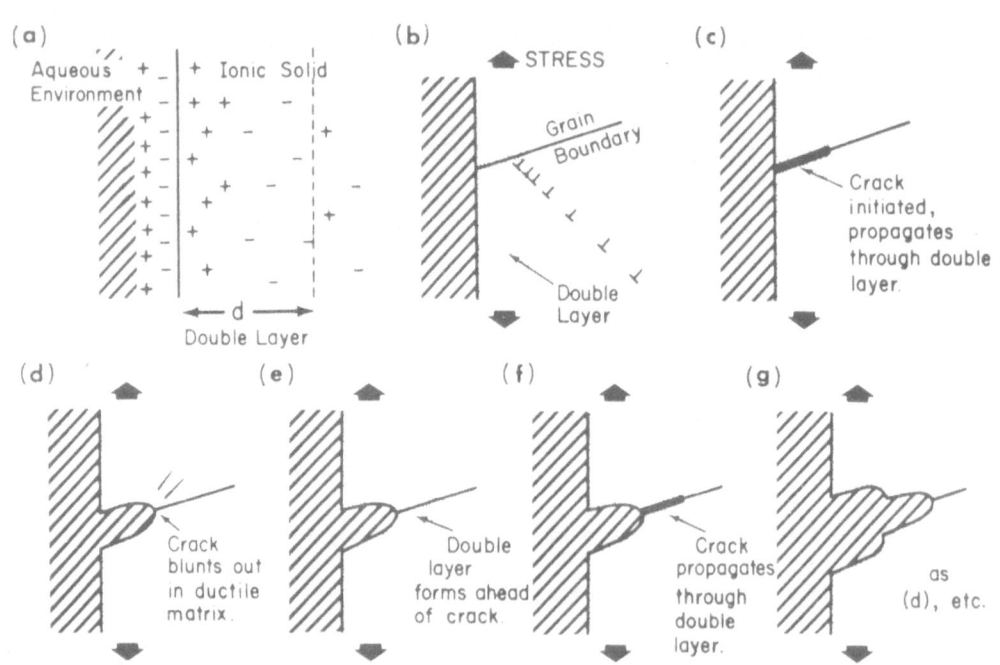

Fig. 10.   Schematic illustration of double-layer mechanism for the complex ion embrittlement of AgCl [37].

certain conditions, this should be controlled by the rate of diffusion of the appropriate charge carriers to the vicinity of the cracktip, and thus be temperature dependent. In particular, the actual relationship with absolute temperature T should depend on the sign of the charge on the adsorbing species. When negatively charged species adsorb, the rate of diffusion of the positive charge carriers, namely, the highly mobile Ag_i^+ ions, should determine the time to failure. For positively charged complexes, the rate of diffusion of the negative charge carriers, namely, the less mobile Ag_v^- defects, should determine t_F. Studies of the temperature dependence of t_F at a given stress confirm this prediction, establishing linear relationships between log t_F and $1/T$ over the temperature range $10-100^{\circ}C$ for specimens tested in solutions containing either negative or positive complex ions (Fig. 11)[37]. In addition, calculations of activation energies from these data yield values in good agreement with those for the motion of Ag_i^+ and Ag_v^- defects determined by other workers.

C.   Control of Complex-Ion Embrittlement

Given that the severity of embrittlement is related to the charge or charge density on the adsorbing complex ions, it is then feasible from a survey of the chemical literature to devise effective counter measures[40]. For example, the successive replacement of sodium ions in 6N aq. NaCl solutions by either potassium or cesium ions inhibits embrittlement, as a consequence of the formation of mixed ions of lower charge and charge density, such as $[Cs^+(AgCl_4^{3-})]^{2-}$, Fig. 12[41]. Likewise, the addition of Group III B cations such as Zn^{2+}, Cd^{2+} or Hg^{2+} to 6N aq. NaCl solutions also inhibits embrittlement, in this case because competition of these ions for chloride ions again results in the formation of complexes of charge less than 3-, for example,

$$ZnCl_4^{2-} + 2AgCl_2^- \rightleftharpoons Zn^{2+} + 2\,AgCl_4^{3-} \ .$$

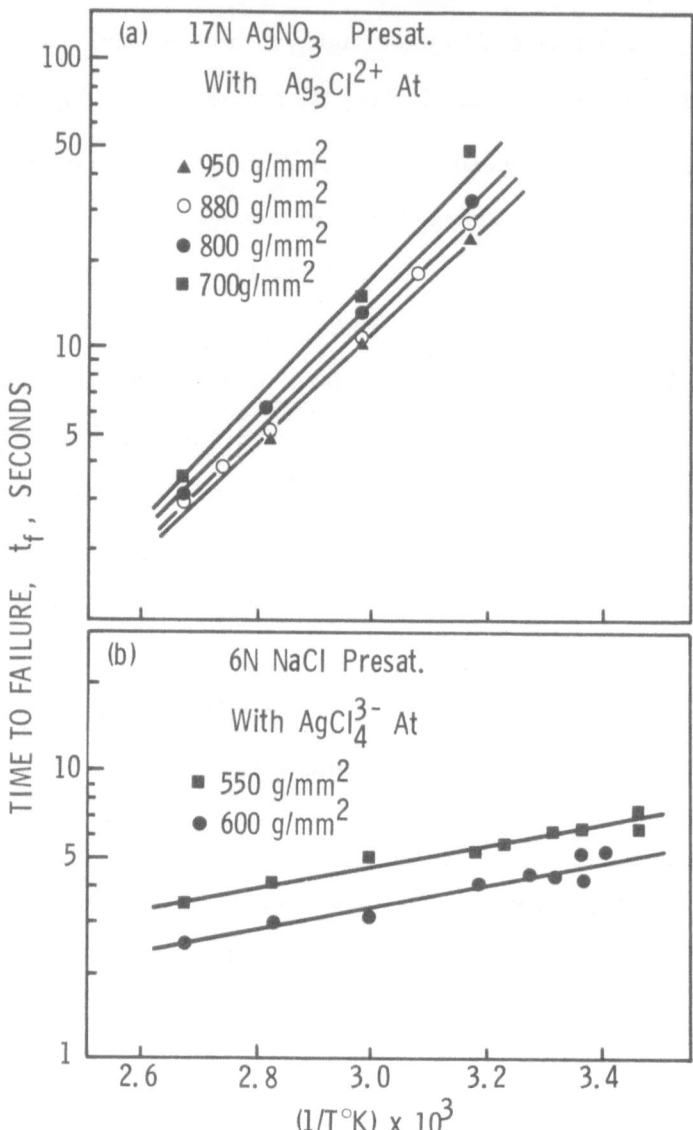

Fig. 11.   Relationship between time to failure and temperature for polycrystalline AgCl tested at various stresses in aqueous solutions containing (a) positive, and (b) negative complex ions [37].

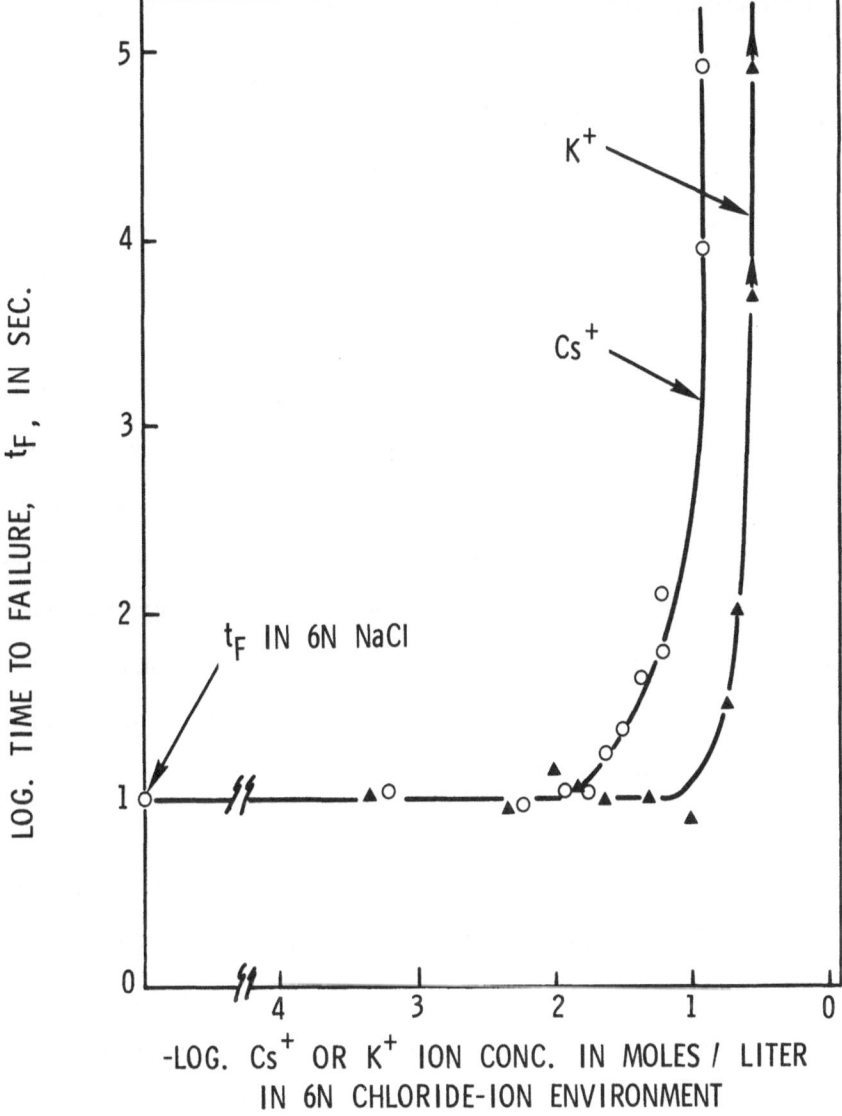

Fig. 12.   Inhibition of embrittlement of AgCl in 6N. NaCl environment by replacement of Na ions by K or Cs ions.

4. CHEMOMECHANICAL EFFECTS IN CERAMIC SOLIDS

A.   Occurrence and Mechanisms

Most ceramic materials are notch brittle and do not possess suf-

ficient slip systems to satisfy Von Mises' criterion for inherent ductility,

Consequently, they usually fracture as the result of dislocations piling up against some stable obstacle after only limited flow has occurred, with the freshly nucleated cracks propagating rapidly to cause catastrophic failure. For this reason, the fracture stress of such solids is determined by, and approximately equal to, their flow stress. An important consequence of this is that the fracture behavior of most ionic-ceramic solids is quite dependent upon the effects of the environment on their near-surface flow behavior. Two other important generalizations may also be made at this point[11]: (i) environments which facilitate flow in the near-surface regions of a notch-brittle nonmetallic solid, i.e., soften it, reduce its fracture strength -- and vice versa, and (ii) environments which markedly influence hardness usually have no significant influence on the energy to propagate cleavage cracks in conventional double cantilever tests, i.e., on the brittle cleavage surface energy of the solid. This implies that the mechanisms by which environment-sensitive fracture occur do not, as proposed by Rebinder[43] primarily involve any adsorption-induced reduction in the surface free energy of the solid. Rather, it is now thought[11, 12, 43] that such effects are associated with adsorption-induced variations in flow (dislocation) behavior, either in the near-surface regions preceding crack initiation or in the vicinity of the crack tip during its sub-critical, slow-growth phase.

Recent studies have also revealed the existence of a most useful correlation between the environment-sensitive hardness and the ζ-potential of a non-metallic solid in a liquid environment. This is that hardness H is greatest when $\zeta \simeq 0$ [44]. Such a "ζ-correlation" has now been established for alumina, calcite, quartz, magnesium oxide, and soda-lime (s.l.) glass in a variety of aqueous and non-aqueous solutions[11]. Thus, it seems likely that this correlation is a property generic to inorganic non-metals, both crystalline and non-crystalline.

Now for crystalline solids, hardness is controlled primarily by the

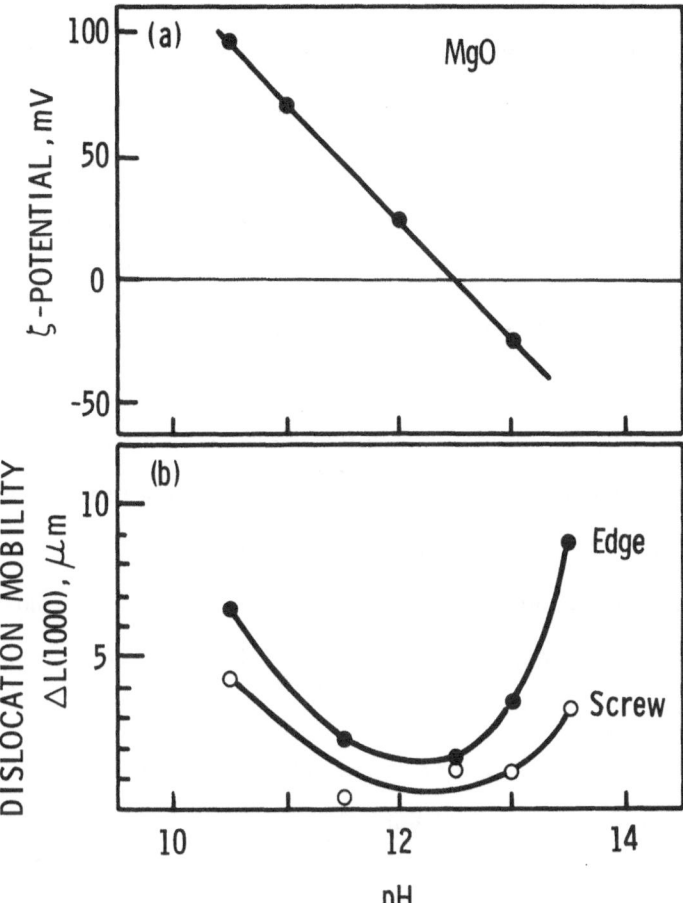

Fig. 13. Variations with pH of (a) ζ-potential, and (b) dislocation
mobility parameter ΔL (1000) for MgO in 10^{-2} N. aq.
NaOH [45]. Note that dislocation mobility is least, i.e.,
the crystal is hardest, when $\zeta \simeq 0$.

mobility of near-surface dislocations. The ζ-correlation implies, therefore,

that such mobility should be a minimum when $\zeta = 0$. Etch-pitting studies

of the movement of edge and screw dislocations around hardness indentations

in freshly cleaved MgO surfaces confirm this expectation, Figs. 13(a) and

(b) [45].

The explanation for the existence of such a correlation remains

obscure at this time, because no detailed mechanistic understanding of the

influence of environment on near-surface dislocation mobility (or flow behavior in amorphous materials) yet exists, nor is any clear understanding available of the correlations between ζ-potential, surface-charge, and near-surface electronic structure. Current thinking[45] is that chemisorption induces a change in the electrostatic potential of the near-surface region of the solid which in turn causes a localized redistribution of the charge carriers. Now, for most ionic and covalent crystals, the carriers involved are electrons, and it is envisaged that their redistribution changes both the band structure of the lattice and the electron occupation of dislocations and intrinsic and extrinsic point defects in the near-surface region. Consequently, the electrostatic interactions between moving near-surface dislocations, between dislocations and point defects, and between dislocations and the lattice, are changed. And since these are the factors that control dislocation mobility in ionic and covalent materials, the near-surface hardness of such crystals is environment-sensitive.

Although surface-active environments do not affect the propagation of fast moving cracks in ceramic solids[46], it seems not unreasonable to expect that they will influence the propagatability of slowly moving cracks by affecting dislocation mobility in the vicinity of the crack tip. Such an effect could either facilitate sub-critical crack growth by dislocation emission processes, leading to delayed failure, or result in crack blunting and reduced propagatability. This should be an interesting and possibly important topic for study, being directly relevant to the anticipated future use of ultrahard, notch-brittle solids as structural materials.

Now if, as suggested above, cracks are initiated in a notch-brittle solid soon after the onset of macroscopic flow, then there should be some correlation between the effects of a given environment on the flow behavior and the fracture behavior of such a material. Such a correlation has been revealed during studies of the environment-sensitive drilling behavior of MgO and CaF_2[47]. One direct measure of dislocation mobility (inversely related to flow stress) is the extent to which dislocations move away from a standard microhardness impression in a given time, say 1000 sec, as revealed by etch pitting. Likewise, one indication of the fracture behavior of a solid, albeit a rather complicated one, is the extent to which a carbide spade bit penetrates the solid under given conditions of thrust, rate of rotation and time. And Fig. 14 indicates that a simple correlation exists between environment-sensitive dislocation mobility and bit penetration for MgO in binary dimethyl formamide - dimethyl sulfoxide (DMF-DMSO) solutions[47]. Specifically, spade bit penetration is greatest for those environments which maximize dislocation mobility and minimize hardness, i.e., presumably, give rise to $\zeta \gg$ or $\ll 0$. A similar correlation has also been noted for CaF_2[47]. Thus, a simple relationship exists between environment-sensitive dislocation behavior and machining efficiency in these ideal cases.

B.   Control and Application of Chemomechanical Effects

Carbide spade bits blunt quickly in hard and brittle solids such as alumina, and so are impractical to use. For such solids, therefore, various types of diamond-loaded bits are usually employed. In this case, an opposite correlation between hardness and drilling efficiency is found[7], namely, that drilling rate is greatest when hardness is a maximum, i.e., when $\zeta \simeq 0$, Fig. 15. Of especial interest is the significant increase in penetration rate for those environments for which $\zeta \simeq 0$ (pentyl and octyl alcohols) compared to water, for which $\zeta \gg 0$. In the particular data shown, the average rate of penetration after 200 sec, D (200), is eight to ten times greater for these two alcohols than for water.

By changing the cutting action of the bit, opposite effects on drilling efficiency can be produced for the same solid by the same environment, i.e., in the presence of the same impurities, and with the same cooling, lubricating and dissolution properties. Such observations demonstrate that

252

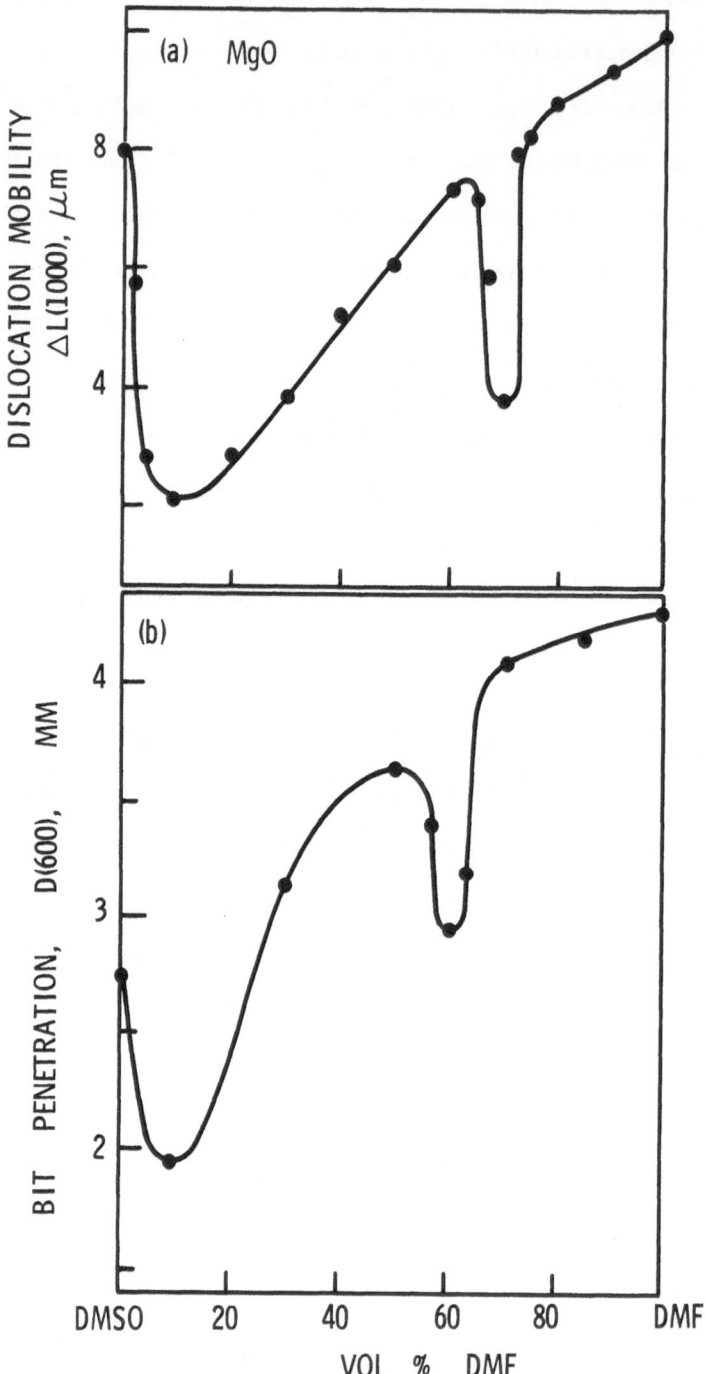

Fig. 14. Variation in (a) edge dislocation mobility [43] and (b) penetration by a carbide spade bit in 600 sec [47] for MgO in DMSO-DMF solutions.

none of the latter factors exerts the controlling influence on drilling rate, i.e., fracture behavior. Rather it is considered that such opposing influences of a specific environment derive from the different roles played by environment-sensitive near-surface flow behavior in the formation of chips by different types of bits[49].

For spade bits, it seems reasonable to suppose that a significant amount of plastic deformation necessarily occurs in a "flow-zone" immediately ahead of the cutting edge, in a manner analogous to conventional metal cutting. Since material in this zone cannot readily escape around the edge of the bit, strain accumulates, rapidly exhausting the limited work-hardening capacity of materials such as MgO or calcite. As deformation proceeds, dislocations pile up at slip band intersections and nucleate cracks. These grow quickly to critical size, and then propagate and interact to complete the process of chip formation. According to this view, then, the rate of penetration of a spade bit varies with environment in the same manner as dislocation mobility, because the dislocation motion that comprises the essential first step in the chip formation process is environment-sensitive.

In contrast, the many irregularly shaped diamonds protruding from a core or hemispherical-ended bit may be regarded as individual cutting tools, each having a short curved cutting edge, a large negative rake angle, and travelling in an essentially concentric circular groove. To the extent that environment-sensitive dislocation motion occurs adjacent to such a tool, it is thought to produce an outward, radial flow of material towards the edges of the groove made by the tool where no further plastic strain can accumulate. Hence, such flow is not envisaged as the primary mechanism of chip formation. Rather, it is postulated that chips are produced mostly by the coalescence of cracks formed immediately behind the tool, where large tensile stresses may be expected in the near-surface region just damaged by passage of the tool. Such plastic flow as does occur is presumed

254

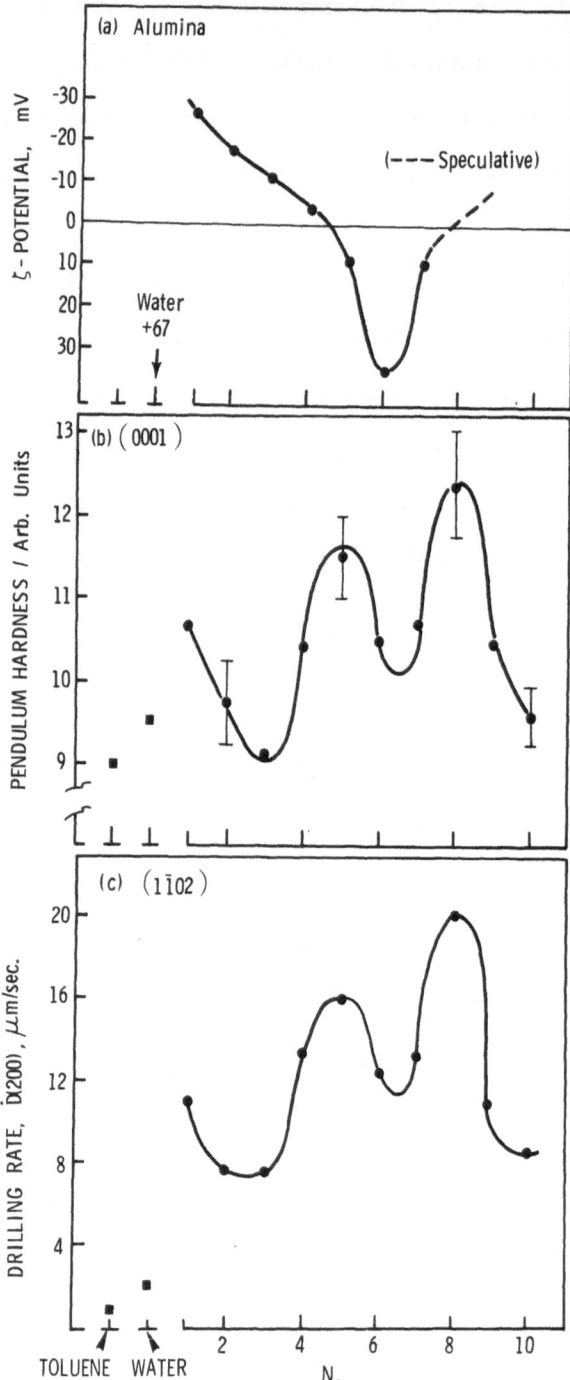

Fig. 15.   (a) ζ-potential, (b) pendulum hardness and (c) rate of
diamond core bit drilling for alumina in toluene, water
and n-alcohol environments [48]. N_C is the number of
carbon atoms in the alcohol molecule.

to both lower the level of stress achieved beneath the tool and to blunt the cracks involved in chip formation -- two negative influences. For diamond bits, therefore, any environment which facilitates dislocation mobility reduces drilling efficiency.

It may be concluded, therefore, that the specific cutting action of the tool, the deformation characteristics of the workpiece, and the influence of the environment on the near-surface flow and fracture behavior of the workpiece must be considered in toto before any recommendation for chemical optimization of some particular machining operation can be made. Extensive testing has revealed that the maxima and minima shown in Fig. 15(c) are quite reproducible for any given type of alumina and testing procedure. Even more surprising, however, is the fact that any particular maximum -- say that occurring in pentyl alcohol -- can be reproduced by mixing together alcohols from the same homologous series on either side of the alcohol in question. Such behavior is illustrated in the data of Fig. 16, and similar behavior has also been observed in drilling studies with glass and granite. Such data establish that the extremum values are real, and not specific to any particular environment. This "mixed-alcohol effect" can be readily understood in terms of the relationship between ζ-potential and hardness. If one component of the binary solution (e.g., hexyl alcohol in Fig. 15(a)) imparts a positive ζ-potential to the solid, and the other (e.g., butyl alcohol in Fig. 15 (a)) imparts a negative ζ-potential, then some mixture of the two components must give $\zeta = 0$, and so produce the appropriate maximum in hardness and diamond bit penetration rate.

This result has extremely useful practical implications for drilling and machining technology. It implies that any liquid which imparts the same ζ-potential to the solid to be drilled will provide essentially the same drilling rate. Thus, a wide choice of cutting fluids is likely to be available for each substance, and it should be possible to find or formulate one that

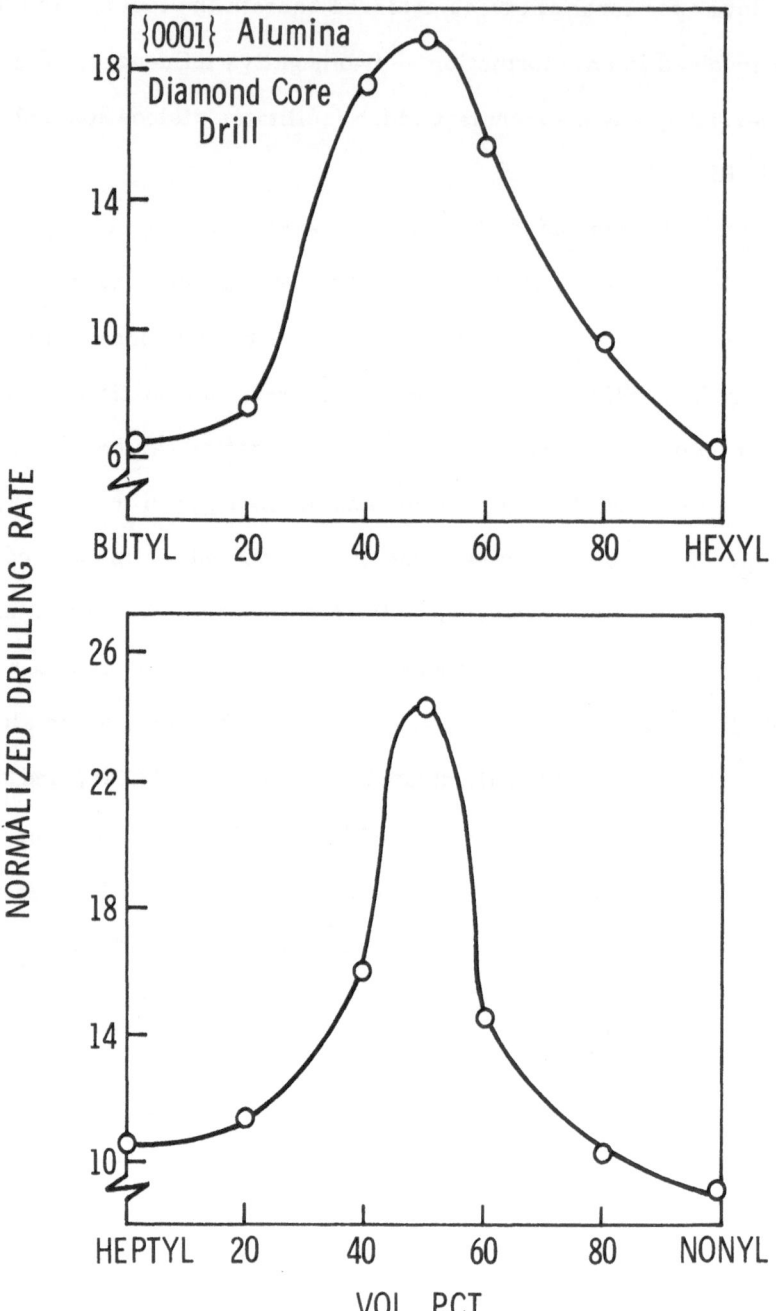

Fig. 16. Rate of drilling of alumina monocrystals in binary solutions of n-alcohols [48].

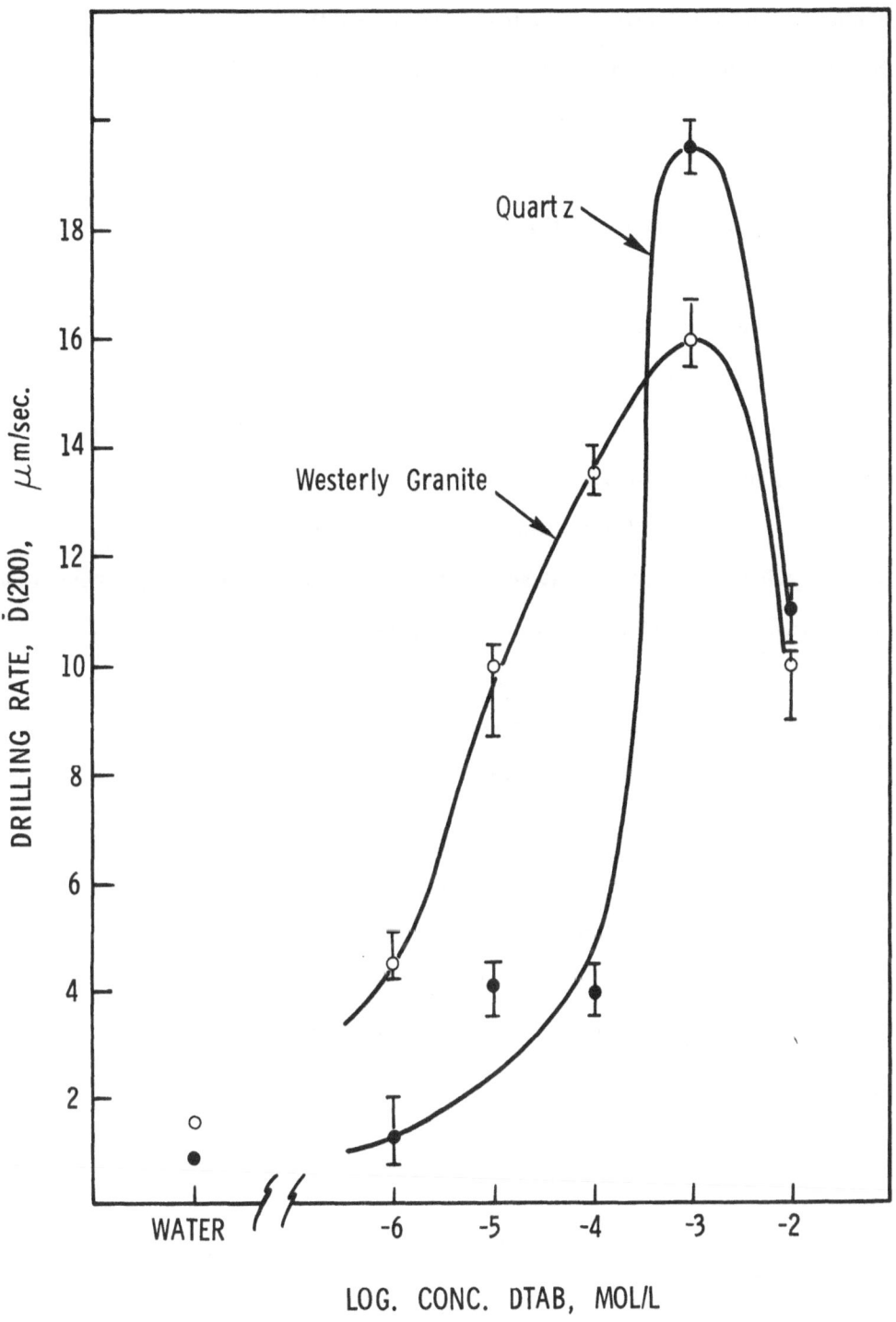

Fig. 17.   Rate of drilling of quartz monocrystals and Westerly
             granite as a function of concentration of aq. DTAB
             environments [50].

is cost effective as well as non-toxic and non-polluting. With this in mind, the concepts discussed above have recently been applied to the laboratory-scale rotary diamond drilling of quartz and several types of granite.

For quartz and Westerly granite, in particular, streaming potential studies indicate that the addition of 10^{-3} - 10^{-4} moles/l of the cationic flotation agent DTAB (dodecyl trimethyl ammonium bromide) to water increase normally negative ζ-potentials of these materials to around zero. Thus, aqueous DTAB solutions of such concentrations might be expected to significantly increase the rates of rotary diamond drilling in these materials, and such is the case, Fig. 17[50]. Note that both quartz and granite show a qualitatively similar variation of drilling rate with DTAB concentration, with maxima at some concentration close to that corresponding to the iso-electric point. It may be concluded, therefore, that the harder quartz phase controls the rotary diamond drilling behavior of granite in aqueous DTAB environments. Note also the eight-fold increase in drilling rate in 10^{-3} molar DTAB over that in water. Other work[51] indicates that up to three-fold improvements can be obtained when the thrusts and rates of bit rotation are adjusted to provide penetration rates similar to those currently used in mole operations in hard rock, i.e., $\sim 10^{-4}$ M/sec (~ 1 ft/hr Bearing in mind that increasing the overall rate of cutter advance by only 25% could result in an 8% reduction in tunneling costs[52], these developments are quite exciting. Similar effects have also been produced in studies involving the use of impact bits[53]. For a more extensive account of the applications of chemomechanical effects, see Refs. 11 and 54.

5. ADSORPTION SENSITIVE FRACTURE BEHAVIOR IN SODA LIME GLASSES

A. Occurrence and Mechanisms

The fracture behavior of a silica glass can be markedly influenced by its testing environment, and both crack initiation[55] and crack propagation[56] processes can be affected. Explanations for the technologically important manifestation of such effects, delayed failure or static fatigue, usually fall into one of two categories, (i) those dependent upon adsorption-induced reductions in surface free energy, and (ii) those which assume the presence of water to be critical[57], and of which the stress-enhanced corrosion mechanism of Charles and Hillig[58] is perhaps most widely accepted. For explanations in category (ii), the role of testing media other than water is usually considered to be simply that of screening the highly reactive water molecules from the glass surface[59].

However, there is much evidence that, under conditions where fracture is induced by abrasion, grinding or drilling, certain organic environments are considerably _more_ active than water[60, 61]. For such examples of environment-sensitive fracture behavior it would certainly appear that explanations dependent upon the dominant presence and corrosive influence of water could not be relevant. Moreover, the same correlation between hardness and environment-dependent surface charge is observed for soda lime glass and fused silica (i.e., hardness is maximum when $\zeta \simeq 0$) as for crystalline solids such as MgO, alumina, or quartz, cf. Figs. 18(a) and (b).

Current thinking is that at least three mechanisms can be involved in the environment-sensitive fracture of soda-lime (s.l.) glasses, the predominant mechanism depending upon the environment involved and testing conditions. There is no question that water significantly reduces the energy to propagate cracks in s.l. glass, typically from $\sim$ 4-5000 erg/cm^2 to 2000-2500 ergs/cm^2 [57, 62, 63]. Likewise there is much evidence to support the existence and practical significance of moisture-related stress-enhanced corrosion effects[64]. The fact that adsorbed water induces a large, negative surface potential on s.l. glass surfaces, reducing its hardness and thus facilitating the near-surface flow processes which can markedly

260

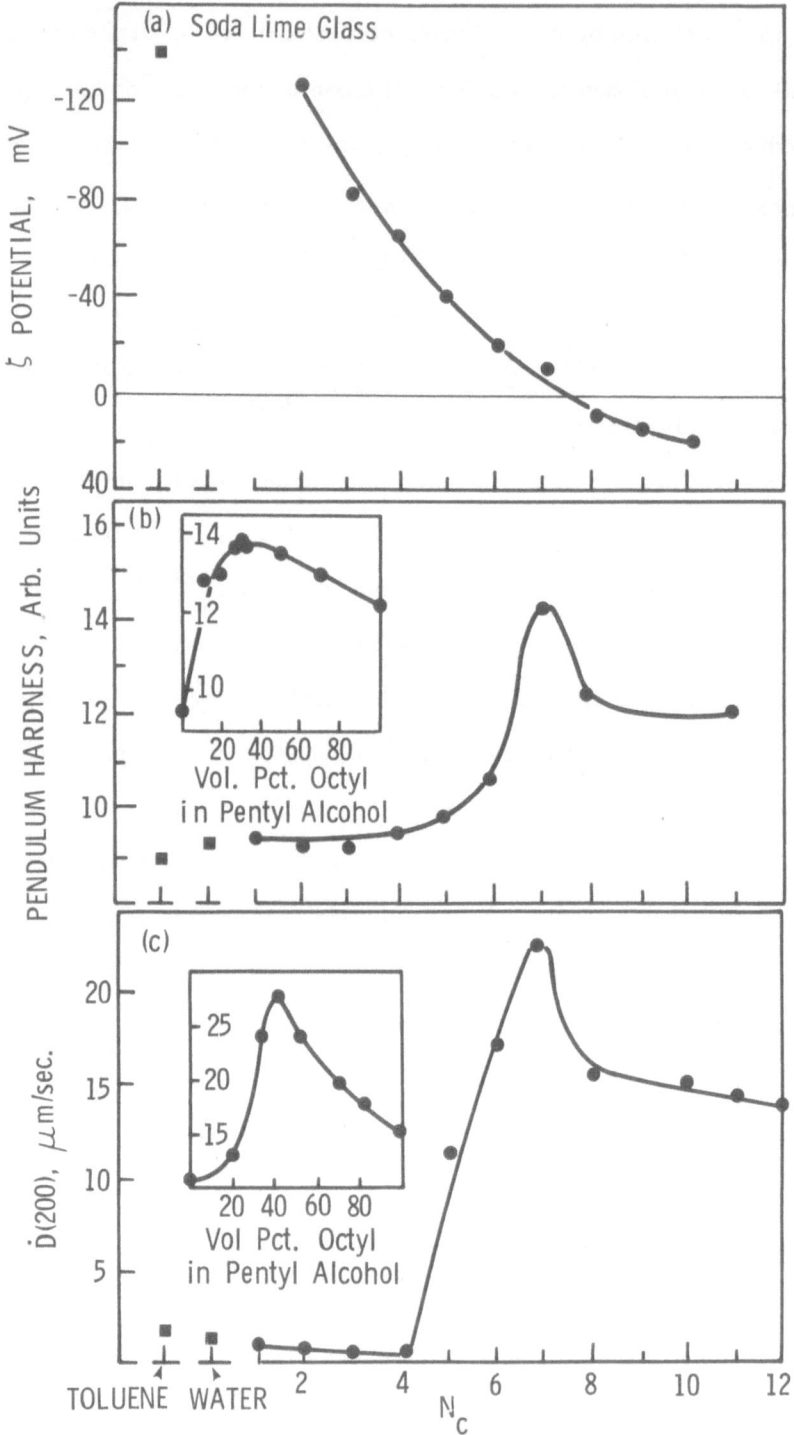

Fig. 18.   Variation in (a) ζ - potential, (b) pendulum hardness and (c)
rate of penetration with a diamond-studded bit of s. l. glass
in toluene, water and n-alcohol environments [49].

influence sub-critical crack growth behavior[8], is less well known,

however, in part because the existence of such effects can only be unequivo-

cally established when the other possible influences of chemical environ-

ments (surface energy reduction and selective dissolution) have been carefully

screened out.

Surface potential-induced effects in s.l. glasses are likely to be

more relevant to crack-_initiation_ dependent phenomena, such as drilling,

than to crack _propagation_-dependent effects. And the similarities in the

characteristics of such effects in crystalline and glassy solids implies the

possibility of some commonality of mechanism. In s.l. glass, the charge

carriers are mobile, non-network ions, e.g., Na^+, OH^-, O^{2-}, and their

distribution is likely to be changed in response to an induced surface

potential[8]. Presumably, an excess of positive ions will accumulate near

the surface when the adsorbate donates electrons to the glass, and negative

ions will accumulate when the glass donates electrons to the adsorbate.

In either case, however, the outcome will be an excess of non-network

ions in the surficial region and, since such ions weaken glass[65], a

decrease in hardness will result for any potential $\neq 0$, as shown in Fig. 18.

Hence, the hardness maximum when $\zeta \simeq 0$.

Of course, similarities in chemisorption behavior between crystals

and glasses of similar composition are not unexpected, because chemisorp-

tion is a localized process determined largely by the properties of active

sites, and because glasses characteristically retain a considerable degree

of short range order and a continuous network of nearest neighbor bonds of

similar geometry and strength to those occurring in the crystalline forms[66]

Furthermore, if flow in glasses proceeds via "dislocation-like" defects,

as suggested recently[67, 68], then (localized) similarities may exist also

in the mechanisms of flow in crystals and glasses. Because of the lack of

long-range order in glasses, however, it is presumably at the atomic scale

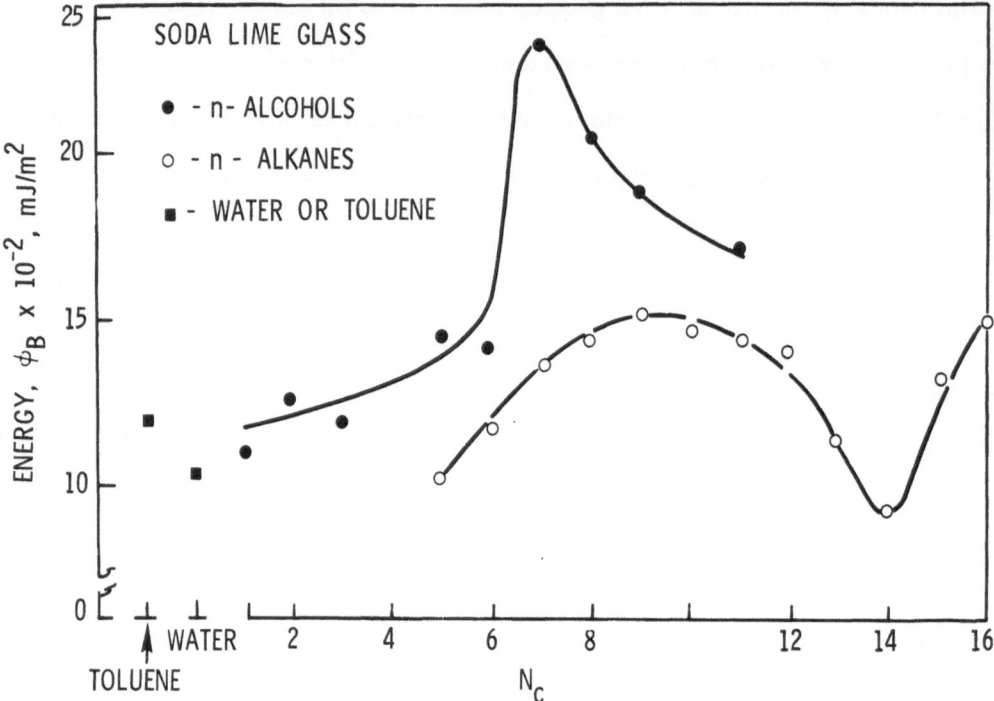

Fig. 19   Influence of environment on energy (ϕ_B) to propagate a
             stable crack in a semi-brittle manner in soda lime glass [8].

that any commonalities of mechanism must be sought. With this in mind,
it has been conjectured[45] that the similarities in the environment-
sensitive mechanical behavior of crystals and glasses arise because of
the similar influence of chemisorption-induced changes in near-surface
electronic structure on the fundamental unit of plastic flow, namely the
formation and/or unit motion of <u>kinks</u> along near-surface dislocations in
crystals or "dislocation-like" defects in glasses.

B.   Adsorption-Dependent Fracture in S. L. Glass

Recent studies using a center-loaded crack technique have led to
the observation of what appear to be adsorption-controlled, flow-facilitated,
slow crack growth effects in soda-lime glass when stressed in non-corrosive
environments[8] Of course, the suggestion that localized plastic flow can

occur in the vicinity of crack tips in soda-lime glass is not new[62, 67].

However, in earlier considerations, flow was not thought as a means of

facilitating crack growth, only the converse.

The data shown in Fig. 19 was obtained using the technique illustrated

in Fig. 20[68]. The specific advantage of this technique is that, within the

dimensional restrictions shown in the figure, the load required to propagate

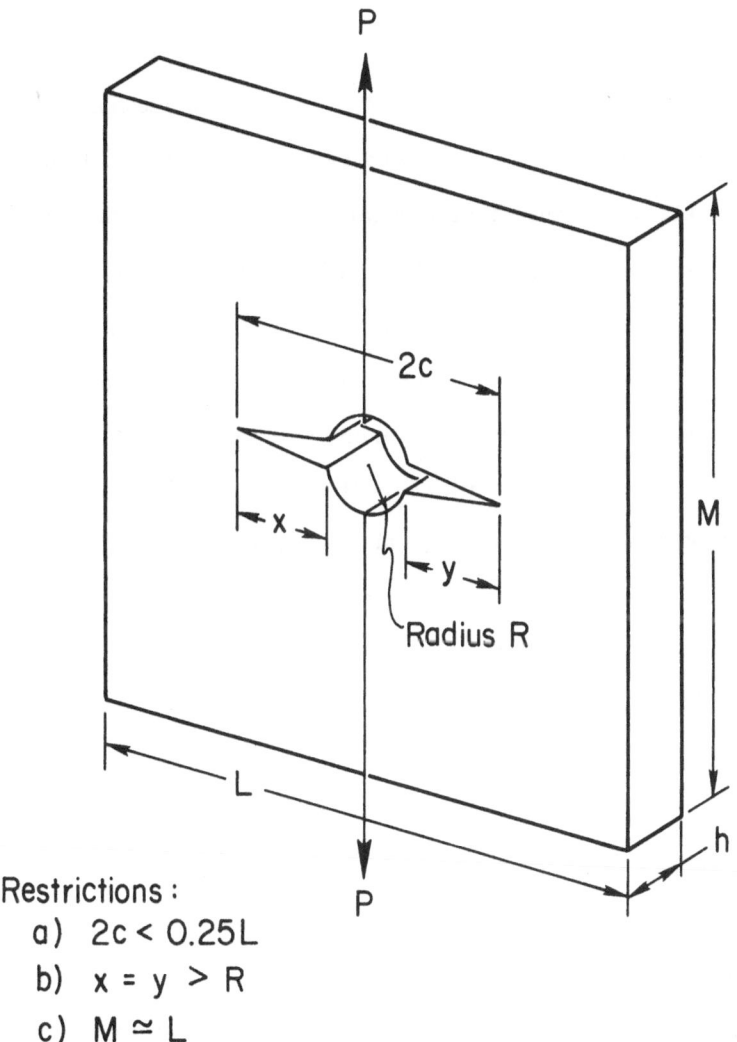

Restrictions:
  a) 2c < 0.25L
  b) x = y > R
  c) M ≃ L

Fig. 20.   Schematic of specimens used in studies of environment-
           sensitive slow crack growth in glass [8].

the crack increases with crack length (the converse of the "Griffith-crack" situation). The crack thus extends in a stable manner at a velocity dependent upon its length and the rate of loading. The energy, ϕ_B, required to propagate center-loaded cracks in a stable mode is calculated from the relationship[69]

$$\phi_B = P^2 (1 - v^2)/2c \pi Eh^2,$$

in which P is the instantaneous load to continue slow propagation of a crack of length 2c, v is Poisson's ratio (0.22), E is Young'e Modulus (71.8 GN/m^2), and h is the thickness of the glass sheet.

Note particularly the maximum in ϕ_B which occurs in heptyl alcohol. This maximum can also be reproduced by solutions of octyl alcohol in pentyl alcohol[8]. Likewise, the maximum (at $N_C \simeq 10$) and minimum at ($N_C \simeq 14$) in the ϕ_B versus N_C (alkanes) plot can be reproduced by, respectively, solutions of tetradecane in pentane and pentadecane in undecane.

Clearly, such behavior is unlikely to be related to any corrosion phenomenon because most of the environments used were, as far as can be ascertained, non-corrosive with respect to glass, and in any event the effects of binary solutions are inconsistent with such a possibility. Nor can such behavior be rationalized in terms of adsorption-induced reductions in the brittle fracture energy of glass for, if this were so, then ϕ_B should be a minimum in heptyl alcohol because, as can be seen from Fig. 18 (b), this environment maximizes the hardness of soda lime glass, and so would be expected to facilitate its failure by brittle processes.

It appears, therefore, that ϕ_B should be regarded _not_ as a measure of the true surface energy of glass, but rather as a relative measure of the energy required to propagate a sub-critical crack in a semi-brittle manner at a rate dictated primarily by the rate of loading[8]. The data can then be interpreted as follows: non-corrosive environments which, by adsorption, soften soda-lime glass (e.g., ethyl alcohol or tetradecane), facilitate slow crack growth by localized flow processes. Therefore, the external load (P) required to maintain crack propagation at a particular rate is reduced, and ϕ_B, which is proportional to P^2, is then relatively low. Environments which maximize the hardness of glass, on the other hand (e.g., heptyl alcohol), minimize flow-assisted crack growth, and in such cases ϕ_B is relatively high.

Such environmentally-induced changes in the hardness of s.l. glass can have profound effects on its machining behavior. The data of Fig. 18(c)[7] show, for example, that under otherwise identical testing conditions, a conventional window glass can be drilled ten to twenty times faster by a diamond-loaded bit in heptyl alcohol than in water. Comparable increases also can be obtained in 10^{-3} molar aqueous solutions of DTAB, which also produce $\zeta \simeq 0$. Appropriately formulated fluids also can increase the rate of grinding of glass by factors of two or three over that obtained in pure water.

6. IN CONCLUSION

Adsorbed, surface active species can influence the fracture behavior of all types of solids, but much more effort of a fundamental and inter-disciplinary nature will be required before the mechanisms involved are understood in any detail. Nevertheless, sufficient insight is now available for a beginning to be made on the control and useful application of their effects. Some examples of the use of adsorption-enhanced fracture be-havior in the drilling of hard solids have been presented, but this is a field in which we have, literally, merely begun to scratch the surface. And the development of highly efficient, economical, water-based, non-toxic and biodegradeable cutting aids for a variety of materials-removing operations represents a real and worthwhile (i.e., cost and energy-conserving) challenge for R&D scientists for the next few years. Likewise, inhibition of the growth to critical size of sub-critical cracks in components by appropriately controlling the chemistry of their operating environments, or perhaps even by adjusting the composition of the solids themselves so that on dissolution they produce preferentially-adsorbed, non-embrittling species, is another area where R&D work could be profitable. In short, the field of adsorption-sensitive fracture behavior is just beginning to open up, and exciting and challenging opportunities are readily apparent for both scientists and development engineers.

ACKNOWLEDGEMENTS

As mentioned in the introduction, this paper is based extensively on earlier review articles published by the author and colleagues. Permission to quote from these works was generously provided by the Journal of Materials Science (for Ref. 11), Plenum Press, Inc. (for Ref. 3) and the McGraw Hill Book Co., Inc. (for Ref. 12). The assistance of C. Joyce Kidwell and R. D. Huntington in the preparation of this contribution is also gratefully acknowledged.

REFERENCES

1. H. K. Huntington, J. Inst. Met., 11, 108 (1914).

2. A. R. C. Westwood, C. M. Preece and M. H. Kamdar, Fracture, Academic Press, New York, 3, 589 (1971).

3. A. R. C. Westwood and R. M. Latanision, Corrosion By Liquid Metals, Plenum Press, New York, 405 (1970).

4. C. M. Preece, Proc. Int. Conf. on Stress Corrosion Cracking and Hydrogen Embrittlement of Iron Base Alloys, Unieux-Firminy, France, June (1973).

5. A. R. C. Westwood, D. L. Goldheim, and E. N. Pugh, Phil. Mag., 15 105 (1967).

6. N. H. Macmillan, R. D. Huntington, and A. R. C. Westwood, J. Matls. Sci., 9, 697 (1974).

7. A. R. C. Westwood and R. M. Latanision, Science of Ceramic Machining and Surface Finishing, Nat. Bureau Stds., Washington, D. C., STP 348, 141 (1972).

8. A. R. C. Westwood and R. D. Huntington, Mechanical Behavior of Materials, Soc. Mat. Sci. Japan, IV, 383 (1972).

9. E. D. Shchukin, A. L. Soshko, O. A. Mitiyuk, and A. N. Tynni, Sov. Matls. Sci., 7, 160 (1973).

10. N. V. Pertsov, E. A. Sinevich, and E. D. Shchukin, Dokl. Akad. Nauk. SSSR, 179, 633 (1968).

11. A. R. C. Westwood, J. Mat. Sci., 9, 1871 (1974).

12. A. R. C. Westwood, C. M. Preece and D. L. Goldheim, Molecular Processes on Solid Surfaces, McGraw Hill, New York, 591 (1969).

13. S. Mostovey and N. N. Breyer, Trans. ASM, 61, 219 (1968).

14. A. R. C. Westwood and M. H. Kamdar, Phil. Mag., 8, 787 (1963).

15. A. R. C. Westwood, C. M. Preece and M. H. Kamdar, Trans. ASM, 60, 763 (1967).

16. W. H. Brattain, The Surface Chemistry of Metals and Semi-conductors, John Wiley, 9 (1959).

17. A. Kelly, W. R. Tyson and A. H. Cottrell, Phil Mag., 15, 567 (1967).

18. T. L. Johnston, R. G. Davies and N. S. Stoloff, Phil. Mag., 12, 305 (1965).

19. M. H. Kamdar and A. R. C. Westwood, Acta Met., 16, 1335 (1968).

20. M. H. Kamdar and A. R. C. Westwood, Phil. Mag., 15, 641 (1967).

21. C.M. Preece and A.R.C. Westwood, Trans. ASM, 62, 418 (1969).

22. N.J. Petch, Phil. Mag., 3, 1089 (1958).

23. E.D. Shchukin, L.A. Kochanova and N.V. Pertsov, Soviet Phys. -Crystallography, 8, 49 (1963).

24. M.H. Kamdar and A.R.C. Westwood, Environment-Sensitive Mechanical Behavior, Gordon and Breach, 581 (1966).

25. R. Rosenberg and I. Cadoff, Fracture of Solids, Interscience, 607 (1963).

26. W. Rostoker, Report No. ARF-B183-12 on Contract No. DA-11-ORD-922-3108, Armour. Res. Fdn., (Nov. 1963).

27. J.V. Rinovatore, J.D. Corrie and J.D. Meakin, Trans. ASM, 61, 321 (1968).

28. W. Rostoker, J.M. McCaughey, and H. Markus, Embrittlement by Liquid Metals, Reinhold (1960).

29. V.I. Likhtman, E.D. Shchukin and P.A. Rebinder, Physicochemical Mechanics of Metals, Academy of Sciences, USSR, Moscow (1962).

30. A.R.C. Westwood, Strengthening Mechanisms: Metals and Ceramics, Syracuse University Press, 407 (1966).

31. Cited in P.A. Rebinder and E.D. Shchukin, Prog. in Surface Science, 3, 97 (1972).

32. C.M. Preece and A.R.C. Westwood, Ref. 3, p. 441.

33. N.S. Stoloff, Surfaces and Interfaces II - Physical and Mechanical Properties, Syracuse University Press, 157 (1968).

34. F.F. Volkenstein, Electronic Theory of Catalysis on Semi-Conductors, Oxford (1963).

35. A. R. C. Westwood, D. L. Goldheim, and E. N. Pugh, Disc. Faraday Soc., 38, 147 (1964).

36. Ibid., Matl. Sci. Res., 3, 553 (1966).

37. Ibid., Phil. Mag., 15, 105 (1967).

38. T. B. Grimley and N. F. Mott, Disc. Faraday Soc., 1, 3 (1947).

39. I. M. Lifshitz and Ya. E. Geguzin, Sov. Phys.-Solid State, 7, 44 (1965)

40. E. N. Pugh and A. R. C. Westwood, Stress Corrosion Testing, ASTM, Philadelphia, STP 425, 228 (1967).

41. A. R. C. Westwood, D. L. Goldheim and E. N. Pugh, Grain Boundaries in Ceramics, Plenum Press, New York, 553 (1966).

42. P. A. Rebinder, L. A. Schreiner and K. F. Zhigach, Hardness Reducers in Rock Drilling, Academy of Sciences, Moscow (1944); transl. C. S. I. R. O. Melbourne (1948).

43. A. R. C. Westwood, D. L. Goldheim and R. G. Lye, Phil. Mag, 16, 505 (1967); 17, 951 (1968).

44. A. R. C. Westwood and N. H. Macmillan, Sciences of Hardness Testing, ASM, Metals Park, Ohio, 377 (1973).

45. N. H. Macmillan, R. D. Huntington, and A. R. C. Westwood, Phil. Mag., 28, 923 (1973).

46. D. A. Shockey and G. W. Groves, J. Am. Ceram. Soc., 52, 82 (1969).

47. A. R. C. Westwood and D. L. Goldheim, J. Am. Ceram. Soc. 53, 142 (1970).

48. A. R. C. Westwood, N. H. Macmillan, and R. S. Kalyoncu, J. Am. Ceram. Soc., 56, 258 (1973).

49. N. H. Macmillan and A. R. C. Westwood, Surfaces and Interfaces of Glass and Ceramics, Materials Science Research, Plenum Press, New York, 7, 493 (1974).

50. N. H. Macmillan, R. E. Jackson and A. R. C. Westwood, Proc. 15th IRSM Symp. on Rock Mechanics, Sept. 1973, in press.

51. A. R. C. Westwood, N. H. Macmillan and R. S. Kalyoncu, Trans. AIME (Mining), 256, 106 (1974).

52. Quoted in Mosaic, a publication of the National Science Foundation, Washington, D. C., 11, Fall 1973.

53. R. E. Jackson, N. H. Macmillan and A. R. C. Westwood, Adv. in Rock Mechanics, Nat. Acad. Sci., Washington, D. C., II B, 1487 (1974).

54. A. R. C. Westwood, Proc. NATO Conf. on Surface Effects in Crystal Plasticity, Germany, Sept. 1975, to be published.

55. I. I. Kitaigorodski and L. N. Kopytov, Sov. Phys. Doklady, 8, 308 (1963).

56. W. P. Berdennikov, Zhur. Fiz. Khim., 5, 358 (1934).

57. S. M. Weiderhorn, J. Am. Ceram. Soc., 52, 99 (1969).

58. R. J. Charles and W. B. Hillig, Symp. on Mechanical Strength of Glass and Ways of Improving It. Union Scientifique du Verre, Charleroi, 511 (1962).

59. R. E. Mould, J. Am. Ceram. Soc., 44, 481 (1961).

60. W. V. Engelhardt, Naturwiss., 33, 195 (1946).

61. R. Ramsauer, Glastech. Ber., 24, 239 (1951).

62. K. R. Linger and D. G. Holloway, Phil. Mag., 18, 1269 (1968).

63. A.R.C. Westwood, G.H. Parr and R.M. Latanision, _Amorphous Materials_, John Wiley, London, 153 (1972).

64. S.M. Weiderhorn, _J. Am. Ceram. Soc._, _55_, 81 (1972).

65. S.M. Cox, _Phys. Chem. Glasses_, _10_, 226 (1969).

66. J.H. Konnert, J. Karle and G.A. Ferguson, _Science_, _179_, 177 (1973).

67. D.M. Marsh, _Proc. Roy. Soc._, _A279_, 420 (1964).

68. V.V. Panasyuk and S.E. Kovchik, _Sov. Phys. Doklady_, _7_, 835 (1963).

69. G.I. Barenblatt, _Adv. Appl. Mech._, _7_, 55 (1963).